Algorithms in Planning 2025

Proceedings

TU Wien
Institute of Architectural Sciences
Research Unit of Digital Architecture and Planning

23 June 2025
Vienna

Edited by
Gabriel Wurzer
Wolfgang E. Lorenz

Cover Design
Gabriel Wurzer and Wolfgang E. Lorenz, based on Penrose Tiling Patterns

Editors
Gabriel Wurzer and Wolfgang E. Lorenz, www.dap.tuwien.ac.at
TU Wien, Institute of Architectural Sciences, Research Unit of Digital Architecture and Planning

ISBN 978-3-384-62543-4

Kontaktadresse nach EU-Produktsicherheitsverordnung
gabriel.wurzer@tuwien.ac.at

Contact address according to EU product safety regulation
gabriel.wurzer@tuwien.ac.at

Print and distribution on behalf of the authors
tredition GmbH, Heinz-Beusen-Stieg 5, 22926 Ahrensburg, Deutschland

Algorithms in Planning 2025

Editorial

Gabriel Wurzer and Wolfgang E. Lorenz
TU Wien – Institute of Architectural Sciences,
Research Unit of Digital Architecture and Planning

Research-driven education ("Forschungsgeleitete Lehre") is common throughout TU Wien, be it in the form of seminars, labs or lectures. As such, the course "Algorithms in Planning" (259.669 SE, 3 ECTS) was nothing special: We looked at recent papers, did some background research on the State of the Art in Algorithmic Planning and compiled these results into written reports (i.e. papers). However, in contrast to regular hand-ins we used a conference system and compiled a proceedings instead of letting the good work go to waste. We also organized a (student-led) conference instead of an end presentation, so this is indeed special and not something that happens regularly at our university. A big "Thank You" to the Dean's office for giving us the Conference Room of the Faculty of Architecture, in which the conference was held on 23rd June between 9am and 3pm. Furthermore, we wish to thank the Head of our Research Unit, Prof. Michael Hensel, for agreeing to hold a Keynote, and, last but not least, all participants of the course for making the event possible:

Aleksic, Lucija
Assmayr, Amélie Elisabeth
Auffarth, Raphael
Aydar, Günes
Destoop, Ellen
Feiziazar, Sama
Grawatsch-Pollhammer, Philip
Großberger, Elisabeth Cäcilia Felicitas
Hochleitner, Antonia Alice
Hoda, Ylleza
Ilker, Sinan
Jakupi, Dorina
Kafashi, Arman
Lim, Hayoung
Müllner, Laura
Neuböck, Nils Frederik
Öksüz, Bükre
Pereira do Carmo, Ângelo Miguel
Ramadani, Besfort
Scheipner, Paul
Seifert, Isabelle
Softic, Emir
Sole, Lucia
Sredzinska, Kinga
Tran, Viet Loi
Weltken, Hanna
Weyrich, Jeff
Yang, Yucheng
Zupa, Lucija

Content-wise, this proceedings book contains the works of our 29 students in the following fashion:

- **Overview chapters** were edited jointly by teams of (ideally) 2-3 students, in order to learn scientific collaboration hands-on. As source, students used a wide variety of scientific indices such as CuminCAD, google scholar, citeseer and so forth.
- **In-depth chapters** giving the State-of-the-Art on a recent topic were compiled by each student, taking recent conferences in their field; for Building Science, we used the proceedings of BAUSIM (available at IBPSA), for Architecture, we used the eCAADe conference (papers available in CuminCAD), for Data Science, we used ieeeexplore.

The papers were entered in OpenConf and automatically compiled into a book using the ProceeDings™ /lite/ system which was developed in-house at the research unit.

Gabriel Wurzer and Wolfgang E. Lorenz
Vienna, June 2025

ORGANIZATION

TU WIEN

dap digital architecture & planning

SPONSORS

ProceeDings

ProceeDings™ /lite/ v1.5 compiler
"In cloud we serve"

Contents

OVERVIEW PAPERS

A Brief Characterization of Planning

Ellen Destoop[1], Sinan İlker[2], Viet Loi Tran[3]
[1,2,3]Technische Universität Wien
[1]e12420000@student.tuwien.ac.at [2]e12432842@student.tuwien.ac.at
[3]e12347304@student.tuwien.ac.at

Planning is a fundamental process that has shaped human environments from ancient times to the digital era. This paper traces the evolution of planning, from early spatial organization in ancient civilizations and the structured methodologies of medieval church construction to modern algorithmic and data-driven approaches.

Keywords: *Evolution of Planning, Strategic Planning, Historical Development, Modern Planning, Algorithmic Planning.*

INTRODUCTION

Planning serves as the foundation for architectural projects, defining the built environment. Fainstein (2025) defines urban planning as a field concerned with the design and regulation of space. It takes into account the physical form, economic functions, and social implications of urban places. Its modern origins can be traced back to 19th-century social movements aiming at improving industrial towns, with early visionaries and practical concerns driving urban planning progress.

Academically, planning theory has expanded to include discussions about techniques, programs, and inclusivity (Fainstein, 2000). While it has been criticized for focusing on method rather than context and outcome, the purpose of planning unavoidably involves the physical city constructed by architects (Fainstein, 2005). From a management standpoint, planning is a major job that entails determining what has to be done, when, where, how, and by whom, a premise that applies to both architectural firms and individual projects (Sarker, 2019).

Planning is critical to architecture since it influences building placement, form, and integration into urban environments (Fainstein, 2025). It seeks to further greater societal objectives such as social equality, economic prosperity, sustainability, and beauty. New Urbanism and other architectural groups advocate for compact, diversified communities with a concentration on public spaces (Fainstein, 2000). Planning also addresses challenges such as urban sprawl and segregation, with architects helping to develop sustainable solutions. Because projects frequently require government clearance for zoning compliance (Fainstein, 2025), architects must grasp planning restrictions. Historically, ideas such as Howard's Garden City and Le Corbusier's Modernism have had a considerable impact on planning, emphasizing the close relationship between the two areas.

The purpose of this chapter is to look at how planning has evolved over time. This investigation tries to demonstrate how planning has constantly adapted to shifting cultural, technological, and environmental needs by tracing its roots back to early practices. This review lays the groundwork for future, more focused discussions in architecture, building science, and urban design.

KEY CONCEPTS & TERMINOLOGIES

Planning in architecture refers to the process of designing and regulating the use of space, focusing particularly on the physical form and spatial organization of buildings, neighborhoods, and urban environments (Fainstein, 2000). It involves a conscious effort to shape the built environment to meet specific goals and needs.

There are some key concepts and terminologies in architecture related to planning according to Fainstein:

- Physical Design: This is a critical component of architectural planning that involves the layout of buildings, infrastructure, and open spaces. This encompasses concepts such as density, architectural typology, and spatial linkages.
- Master Plan: A master plan is a comprehensive long-term plan used in architecture and urban planning to give a conceptual framework for future growth and development. It frequently incorporates zoning restrictions, infrastructural plans, and architectural guidelines.
- New Urbanism: This is a significant architectural and planning movement that advocates for compact, mixed-use neighborhoods with a focus on walkability and a strong public realm. It emphasizes traditional neighborhood design principles.
- Urban Design: This aspect combines architecture and urban planning by focusing on the physical design of cities and towns, such as buildings, streets, parks, and public spaces.
- Zoning: Zoning is a crucial regulatory instrument in urban planning that separates land into districts based on land use, building height, density, and other physical characteristics. Architects must design within these limits.
- Public Realm: The term "public realm" refers to publicly accessible locations inside a city or community, such as streets, parks, and plazas. Architectural design influences the quality and utility of the public environment.
- Built Environment: This refers to all human-made structures and locations, such as buildings, infrastructure, and urban areas. Architectural planning is primarily concerned with shaping the built environment.
- Sustainability: Environmental sustainability principles are increasingly being included into architectural planning, with the goal of designing buildings and urban spaces with as little environmental impact as possible.
- Aesthetic Satisfaction: Architectural planning takes into account the built environment's visual and sensory experiences, with the goal of creating aesthetically pleasing designs.

Furthermore, the multidimensional role of planning in society has profound implications for architecture:

- Shaping Urban Form and Character: Architectural planning plays an important role in creating the physical shape and character of urban areas, influencing anything from development density to architectural styles found across a metropolis. The ideals advanced by movements such as New Urbanism are explicitly intended to modify urban form.

- Addressing Social Needs: Architecture planning may help to address social needs by developing inexpensive housing, making public areas more accessible, and encouraging a sense of community. The "just city" model's emphasis on equity is closely related to how architectural planning may benefit varied communities.
- Economic Development: Architectural planning contributes to economic development by encouraging investment, establishing bustling commercial areas, and promoting tourism. Economic growth relies

heavily on strategic land use and infrastructure planning.

- Environmental Management: It requires sustainable architectural design and urban planning to reduce energy use, manage trash, and preserve natural resources.
- Regulation and Control: Zoning and construction codes serve as a framework for regulating growth while also protecting public safety and well-being. Architects must include these regulations in their designs.
- Public Participation and Inclusivity: Modern planning theories, such as the communicative model, emphasize the significance of incorporating the general public in the planning and design process. This can make architectural projects more sensitive to community demands and preferences.
- Historical and Cultural Context: Architectural planning frequently takes into account a location's historical and cultural context, with the goal of preserving history and creating designs that are sensitive to its surroundings.
- Addressing Urban Challenges: By encouraging more compact, mixed-use, and egalitarian development patterns, architectural design can help to address urban issues including sprawl, segregation, and inadequate infrastructure.

In essence, planning provides the framework within which architecture operates, guiding the design and development of the built environment to serve various societal goals and needs, ranging from the functional and aesthetic aspects of individual buildings to the larger organization and character of cities and regions.

HISTORICAL PERSPECTIVES

The history of planning can be understood as a gradual evolution, shaped by societal needs, and changing ideas about space and organization. Early planning efforts were closely tied to religious and medieval influences, such as the construction of churches. These religious centers were often the points around which towns and cities grew, which reflects an early form of urban planning. Medieval societies also engaged in more rudimentary urban and regional planning to improve trade, defense, and governance. City walls, market squares, and road networks were designed to support economic and social functions.

Renaissance

During the Renaissance and pre-modern era, planning became more structured, influenced by advancements in art, science, and philosophy. During this period, design principles that highlighted balance, symmetry, and functionality became more important. Architects and planners viewed cities as works of art, integrating aesthetic considerations with practical urban challenges. The influence of classical architecture and scientific discovery contributed to a more intentional and rational approach to spatial organization. The Renaissance also introduced the concept of ideal cities, where geometry and proportion were used to create harmonious and efficient urban environments.

Industrial Revolution

The Industrial Revolution marked a turning point in planning history, as rapid urbanization and technological progress transformed cities. Factories, railroads, and mass housing developments reshaped the built environment, but these changes also introduced major challenges, including overcrowding, poor sanitation, and social inequality. In response, early modern planning theories were needed to address these issues. Governments and reformers introduced zoning laws, sanitation policies, and new urban layouts, which were designed to improve living conditions. City planning became essential for managing growth, regulating land use, and ensuring public health. These early efforts laid the foundation for contemporary urban planning, which continues to evolve in response to emerging challenges and new societal needs.

20th century

During the 20th century planning theory was further refined with the influences by figures such as Patrick Geddes, who emphasized the relationship between geography, environment, and planning. Le Corbusier's vision of the "Radiant City" promoted high-density, functionally segregated urban designs, while Frank Lloyd Wright championed decentralized urban forms, advocating for a closer integration with nature. These differing perspectives shaped the debates on how cities should be structured, balancing efficiency with human-centered design.

Post-war

After World War II, planning took on a more modernist approach, focusing on large-scale infrastructure projects such as highways and mass housing developments. However, these top-down approaches often overlooked the social fabric of communities, leading to criticism from urban theorists like Jane Jacobs, who argued for more participatory urban planning that prioritized local engagement and human-scale design.

Present

Planning according to Robert Freestone (2018) is deeply connected to its role as an action-oriented practice, where assessing past decisions and policy outcomes is essential. However, interest in planning history extends beyond practical concerns, driven by diverse intellectual and interdisciplinary influences. Modern planning integrates insights from various fields, reflecting a continuous effort to refine and adapt strategies in response to changing societal needs.

CONTEMPORARY PERSPECTIVES ON PLANNING

The concept of planning represents one of humanity's most fundamental organizing principles, evolving dramatically from its early times in religious architecture to today's complex digital systems. Modern planning has transitioned from rigid, centralized systems to flexible, adaptive frameworks that could integrate diverse disciplines together and with outcomes shaped by rapid global transformations. Contemporary planning not only organizes these physical arrangements of urban spaces but also addresses social, economic, and environmental order.

This section examines the transition to modern planning frameworks, technology's growing influence, and emerging trends reshaping planning that practices globally.

Transition to Modern Planning Paradigms

Historically, planning is often a top-down process driven by experts. Today, however, planning increasingly relies on participatory models that incorporates input from local communities and stakeholders help. Each paradigm shift has typically emerged in response to recognized limits within preceding approaches, making these transitions mainly correct in nature and beneficial to the profession's advancement (Chandy, 2023).

Planners now consider a broader spectrum of factors, including socio-economic differences, environmental sustainability, and cultural dynamics to create adaptive strategies that will meet the challenges of a connected world. It has moved beyond the rigid rational-technical model of the mid-20th century, which was rooted in positive epistemology and modernist traditions. This transition began with the recognition that planning is not merely a technical exercise but fundamentally a socio-political process embedded within specific cultural contexts (Watson, 2016).

Research into contemporary planning paradigms has identified its four fundamental frameworks of strategic importance for sustainable industrial and urban development era:

- The systemic-civilizational paradigm, which reconceptualizes planning within broader civilizational contexts
- The socio-humanitarian paradigm, emphasizing human-centered approaches
- The innovation paradigm, prioritizing creative and adaptive solutions
- The globalization paradigm, accounting for worldwide interconnections and influences (Eren & Kara, 2023).

Influence of Technology and Globalization

Technology and globalization have fundamentally reshaped planning practices across all data. The integration of digital technology and the increasing globalization helped the main points of planning reshaped in its process. The integration of digital tools, big data, and algorithmic methods has enabled the planners to simulate urban dynamics, foresee trends overall, and optimize resource allocation.

The rise of digital technologies, from geographic information systems (GIS) to algorithmic modeling, has transformed planning into much more data driven discipline. Recently, algorithmic urbanist approach and data-driven tools have made planning more adaptive, predictive, and responsive overall. These tools not only enhanced the predictive capabilities but also supported the collaborative decision making in a global context. These technological advances serve as powerful catalysts for reducing global connection barriers and transforming modern society (Barak & Kılınç, 2023).

BROAD TRENDS SHAPING CURRENT PLANNING PRACTICES

Current planning trends are driven by the urgency of sustainable development and climate resilience in the fields. Practices, such as smart cities, which integrate digital technology with urban infrastructure, that illustrate how planners are reimagining traditional models to gain more sustained and livable urban environments in the global planning.

The relationship between mobility and urban planning has always been intimate, with cities growing in response to changing transportation needs. Contemporary planning must now accommodate new modes of transport, including autonomous and electric vehicles. According to recent research, infrastructure for electric vehicles is rapidly expanding, with projections indicating 1.3 million public charging stations by 2025, and 2.9 million by 2030 (AtkinsRéalis, 2022).

SYNTHESIS & FUTURE OUTLOOK

A historical overview of planning shows a continuous evolution—from the spatial organization of early civilizations to the complexities of the modern era (Georgia Planning Association, 2023). The broad trends and transformative approaches outlined in this chapter serve as a foundation for the subsequent explorations in architecture, building science, and urban design. Each of these specialized domains takes up addressing the modern challenges by adapting and extending the principles of contemporary planning.

While contemporary planning has embraced many transformative approaches, the integration of emerging technologies such as artificial intelligence (AI) is poised to redefine the planning process in unprecedented ways. In this context, a key question arises for future practice:

How can planners use data-driven tools and community input to create more sustainable and climate-resilient cities, especially in fast-growing urban areas?

The researchers will focus on 'in what ways can planners better integrate sustainability and resilience in rapidly urbanizing regions'. These open questions will contribute and invite new explorations and innovation in the field, ensuring that planning remains responsive to the evolving needs of society.

REFERENCES

AtkinsRéalis. (2022, July). What are the *trends shaping the future of urban planning?* [Online]. Available at: https://careers.atkinsrealis.com/blogs/2022-7/what-are-the-trends-shaping-the-future-of-urban-planning.

Barak, B., & Kılınç, S. (2023). Paradigm shifts in urban planning: A bibliometric analysis. European Journal of Geography, 14(2), 43–58. [Online]. Available at: https://eurogeojournal.eu/index.php/egj/article/download/212/166/334.

Georgia Planning Association. (2023). History and theories of planning: AICP exam prep. [Online]. Available at: https://georgiaplanning.org/wp-content/uploads/AICP_History-and-Theories-of-Planning_April-2023.pdf.

Chandy, J. (2023). A speculative exploration into the current planning paradigm through academic planner perspectives, University of Waterloo Institutional Repository. [Online]. Available at: https://dspacemainprd01.lib.uwaterloo.ca/server/api/core/bitstreams/1a4b620f-05c8-4d2b-987c-5a3691b2bc29/content.

Eren, M., & Kara, N. (2023). Transformation of planning paradigms: A historical and theoretical perspective. Economy and Policy, 3(2), 126–137. [Online]. Available at: http://ojs.econindustry.org/index.php/ep/article/view/267.

Hackett, B. (1950) The Evolution of Planning Theory: Needed: A Social, Physical, and Economic Philosophy, Landscape Architecture. Washington, D.C: American Society of Landscape Architects, Publication Board, pp. 114–118.

Hein, C. (2018) *The Routledge handbook of planning history*. New York, NY London: Routledge.

Fainstein, S.S. (2000) 'New Directions in Planning Theory', *Urban affairs review* (Thousand Oaks, Calif.), 35(4), pp. 451–478. Available at: https://doi.org/10.1177/107808740003500401.

Fainstein, S.S. (2005) 'Planning Theory and the City', *Journal of planning education and research*, 25(2), pp. 121–130. Available at: https://doi.org/10.1177/0739456X05279275.

Fainstein, S.S. (2025). Urban planning | Definition, History, Examples, Importance, & Facts. *Encyclopedia Britannica*. Available at: https://www.britannica.com/topic/urban-planning.

Sarker, Anupom. (2019). Planning Function of Management.

Watson, V. (2016). Shifting Approaches to Planning Theory: Global North and South School of Architecture, Planning and Geomatics, University of Cape Town.

Planning in Architecture

Leveraging Data-Driven Programs in Architectural Planning

Hayoung Lim
Technical University of Vienna
e12348271@student.tuwien.ac.at

Studying the use of data-driven tools with an emphasis on pre-design and feasibility is part of the exploration of tools used in the early phases of architectural planning. This study categorizes important data types and looks at how these data types are analyzed by tools. The study will demonstrate how tools improve design sustainability and accuracy and how they facilitate well-informed decision-making at the outset of the architectural process.

Keywords: *Architecture planning, Data driven programs*

Introduction

At one point planning in the field of architecture was, to the greatest extent possible, carried out exclusively in accordance with the empirical knowledge of the time, and was limited to the structuring of space. Today, the rapid development of technology has increased both the variety of data that can be collected, as well as the volumes of that data, and has made the data more accurate and reliable. This is because several data-based tools have now been developed and are actively used in the design process. The more quantitative and compound the data that was previously impossible to predict, the more exact its simulation in advance and the more optimal the planning strategies. Therefore, not only the design coherence, but the sustainability of the issue is ensured. Hence, with reference to the modern planning of architecture, the data-driven programs are seen to be significant from the very first stage.

The aim of the given work is to discuss the methods of data-based analysis collecting and utilizing in modern planning in the field of architecture, as well as in design, including a systematic classification of the data, which can be collected at the initial stage of design and a description of the methods of their analysis and use through various digital tools and technological means.

Type of Data Collected in Architectural Planning

Planning in Architecture focuses on a complete and systematic evaluation process, which combines all the information that has been created during the entire life cycle of an architectural project. After the analysis, strategic directions, design guidelines, and ways of action are pre-coordinated. The results of this analysis are the logical basis for making decisions in any phase of project design, construction, and operation. That's why it is important to understand and analyze all the elements involved in a clear and diverse way: functional requirements, spatial context, technical conditions, economic feasibility, sustainability, and user experience.

Before diving into the principal study, this thesis aims to classify architectural planning phases that are deemed necessary for data collection and the kind of data that is required. Consequently, planning is two main phases: Pre-Design & Feasibility Planning and Strategic & Conceptual Planning.

The phase of Pre-Design & Feasibility Planning involves partial research and feasibility studies which occur in the early stages of the architectural project. The aim of the given stage is to assess the feasibility of the project from various angles, to determine the problems of a technical, economic, legal nature to form the initial direction. By doing so, it becomes possible to detect risks in advance, providing a crucial basis for the "Go/No-Go" decision regarding project execution. The specific types of data addressed in this study during the Pre-Design & Feasibility Planning phase are as follows:

- Site Analysis: Examination of geographical characteristics, ground conditions, environmental constraints, and accessibility to infrastructure
- Regulatory Review: Analysis of zoning regulations, floor area ratio and building coverage limitations, and permit requirements
- Energy and Cost Forecasting: Prediction of expected energy consumption, operational costs, and lifecycle analysis of materials

After the initial feasibility review, the Strategic & Conceptual Planning phase follows, which goes beyond just a spatial layout or conceptual idea. It considers the establishment of the terms of reference for the design and execution process. Thus, in this stage, data on social, physical, technical, and environmental factors are collected that are reflected in the design. The information received because of this analytical process is used primarily for making informed decisions about the spatial and architectural arrangement of buildings and related components. The specific types of data addressed during the Strategic & Conceptual Planning phase are as follows:

- Environmental and Energy Performance Analysis: Assessment of sunlight exposure, solar radiation, shading, airflow, and thermal comfort
- Review of Materials and Structural Concepts: Data analysis on structural stability and building materials
- User Experience-Based Spatial Configuration: Analysis and prediction of user experience patterns.

The research analyses data-driven analysis tools with focus on their application during Pre-Design & Feasibility Planning stage.

ArcGIS and PlAXIS

This report has chosen ArcGIS and PLAXIS data-driven applications among many for the analysis. The data-driven programs are programs that use spatial and structural information to assess the feasibility of a design. The report will reveal how these tools work, their operation principles, how they represent information, and how they utilize the information for analysis. ArcGIS is an information system that provides an appropriate basis for determining the site strategies and programming by considering all physical, legal, and environmental conditions of the site. The operation of ArcGIS is based on spatial layering as well as the attribute-based querying.

The spatial layering is a basic structure in GIS. It separates different types of geographic data into layers and represents them in overlapping. You can then analyze the spatial relationships between these layers. Each independent layer that contains a specific theme shares a common coordinate system. So, it creates an integrated analytical environment.

It is possible for users to selectively co-locate only the layers that are relevant to their purpose and visually compare or analyze them with ease. In addition, spatial information can be created based on the user's objective. In attribute-based querying, users can filter information based on the attribute data associated with spatial objects. In ArcGIS, not only the location data but also the attribute data of each spatial object are saved, structured as an attribute table. Users can selectively extract the necessary information by defining certain conditions based on this attribute data.

Applying these principles, ArcGIS becomes an indispensable assistant in examining the legal restrictions applied to a site from a visual perspective. Thus, spatial information, such as zoning, building coverage ratio (BCR), floor area ratio (FAR), height restriction zones, development-restricted areas, and cultural heritage protection zones, can be loaded as individual layers for further analysis. As a result, users can visually see all legal restrictions that overlap with the site and confirm in advance the range and conditions of design feasibility. According to these principles, ArcGIS software is most useful at the early design stage for finding candidate sites that meet certain conditions or analyzing the development potential of certain parts of the city. For example: Extraction of candidate plots for redevelopment, Location analysis that satisfies condition A but excludes condition B, Filtering of development-restricted zones.

PLAXIS is an advanced geotechnical software application that is typically used to determine the structural response of the ground based on site soil and other conditions. It is usually applied in the early stages of architectural design to evaluate the structural risk. Thus, it determines the technical feasibility of a design proposal.

of GIS data PLAXIS core principle essentially is derived from Finite Element Method (FEM) based geotechnical analysis. This is combined with the detailed modeling of soil materials to get a reasonable approximation of true soil behaviors. The Finite Element Method (FEM) is a method to divide the actual problem (which is something unknown) into many small elements. Here, let us split the complex continuous medium (e.g., soil) into multiple small elements. The unknown quantities, i.e., the stress-strain relationships between them, are calculated by way of mathematics, and the overall behavior is analyzed. As for complex structures or materials, first, they are broken down into smaller elements

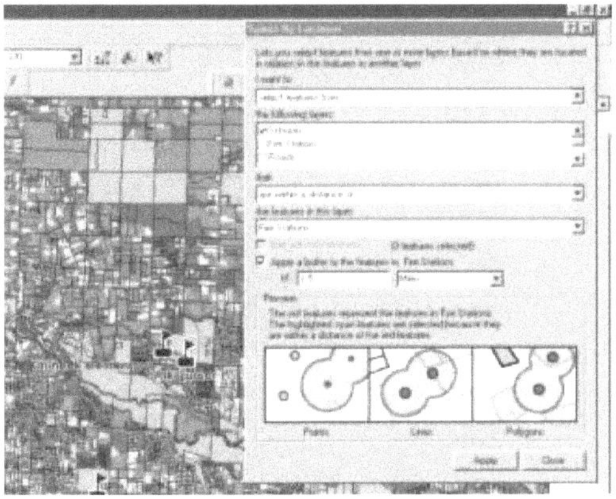

Figure 1 : Relationship among geographic features

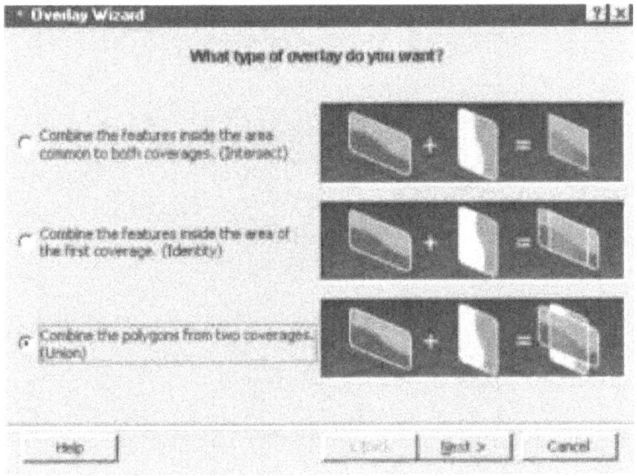

Figure 2 : Overlay Function

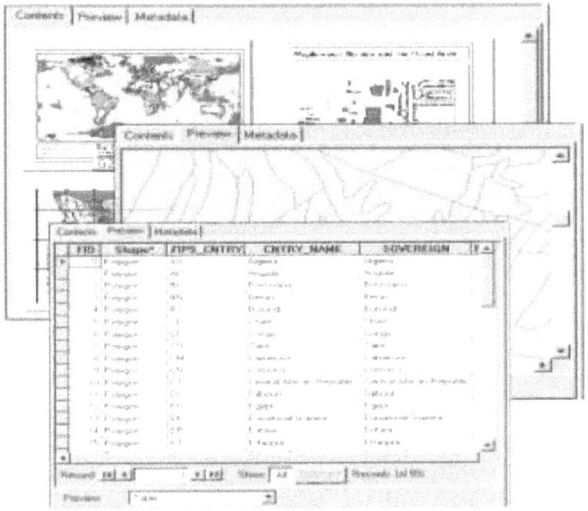

Figure 3 : ArcCatalog Framework for organizing large and diverse stores

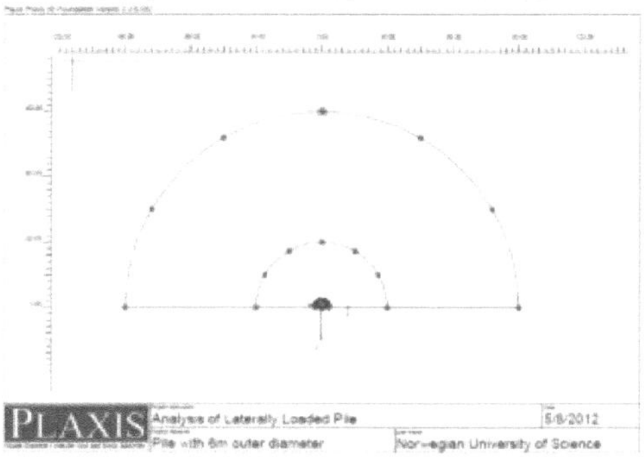

Figure 4 : Defining the Different Cluster for Soil Modeling

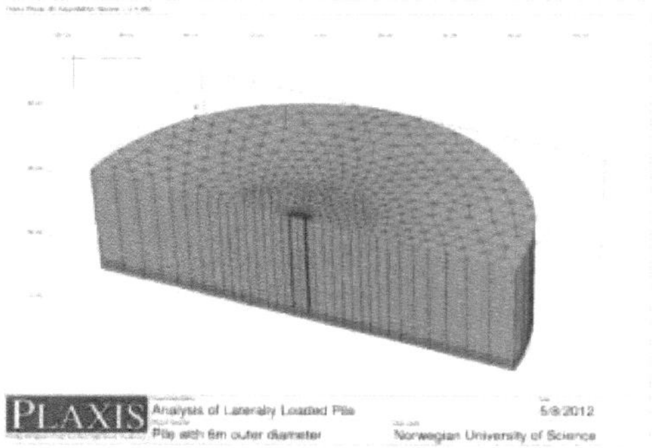

Figure 5 : Three Dimensional Models of the Soil and Pile with Meshing

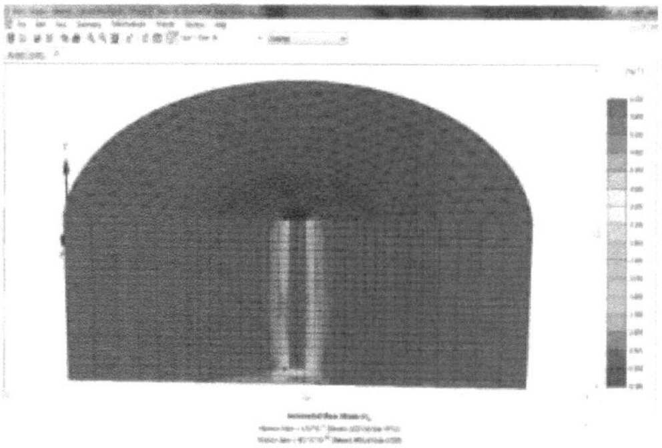

Figure 6 : Incremental Shear Strain

that are simpler to analyze. The internal forces and deformations that occur in each element are calculated and synthesized into a global prediction model. This method is particularly effective for nonlinear and heterogeneous materials, such as soil, and it allows users to quantify complex risks.

PLAXIS. Besides in, PLAXIS meets need different dirt's laws so that users can make a model from site in 2D/3D, input structural load, surface water, underground water, drainage, then simulate the soil respond to the previous loads, to a high precision. As a result of which, simulation's result is not only simple based on modulus but includes the real properties of soil and the property of soil during the loading. The result can go beyond the imagination because the different kinds of soil and level of moisture can be the cause of the differences even if the loading is the same, this feature allows the simulation to be more realistic.

Based on borehole data and laboratory soil testing, users can configure input parameters, anticipate potential risks before construction, and determine suitable technical alternatives. From a long-term perspective, PLAXIS also enables evaluation of whether the planned structure can stably bear structural loads on a given site.

One Click LCA

One Click LCA was chosen for in-depth examination out of all the programs used for energy consumption and cost prediction in this study. A data-driven platform called One Click LCA makes it possible to evaluate a building's economic performance and environmental impact simultaneously. Life Cycle Assessment (LCA), which measures environmental impact indicators like carbon emissions from building materials, equipment, and energy use, and Life Cycle Costing (LCC), which calculates overall costs by taking material lifecycle, maintenance, and replacement cycles into account, are the two main ideas upon which the platform is built. Establishing life cycle-based energy and cost estimation models, mapping to environmental product declarations (EPDs), and automatically extracting material-level data are the fundamental working principles of the program which automatically determines the material name, quantity, specifications, and surface area of each building component when a

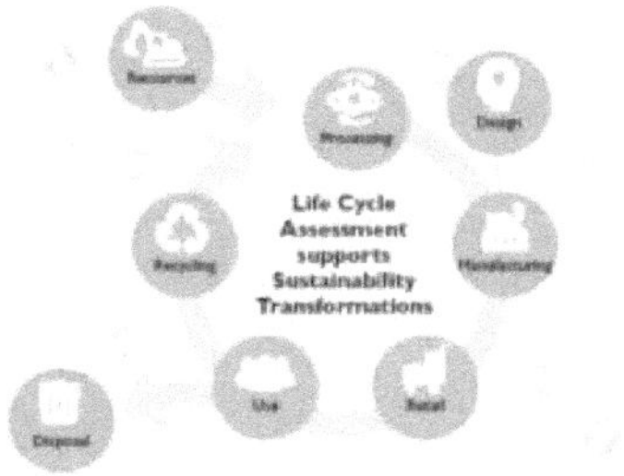

Figure 7 : Product's life cycle

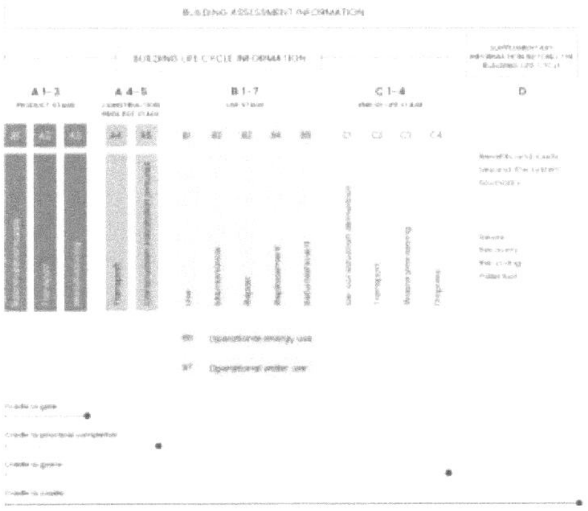

Figure 8 : Buildings' life cycle stages according to EN standards

user enters a BIM model, material specification sheet, or project configuration data. In other words, material data can be directly extracted from user-provided files using One Click LCA. The program uses this information to link pertinent data, including each material's carbon emissions, lifespan, replacement cycle, energy consumption, and maintenance costs. These datasets guarantee practical usability in real-world applications because they are based on environmental product declarations.

One Click LCA runs simulations using a "Cradle to Grave" methodology. This approach considers a material's whole life cycle, including extraction of raw materials, manufacturing, transportation, and construction, as well as use, upkeep, and demolition/disposal. Final evaluation values are generated by quantifying and adding up the costs and environmental impacts (such as carbon emissions and energy consumption) for each period. These simulations allow for the automatic computation of energy consumption and overall cost over the building's service life by accounting for material lifespan, maintenance schedules, and replacement timing. These analyses' outcomes are automatically produced as reports. As a result, users can choose the most sensible and sustainable design option by simulating and comparing the energy consumption and operating costs of various design alternatives.

Climate Studio

Climate Studio was chosen for analysis in this study out of all the data-driven programs used to collect environmental data. A simulation tool called Climate Studio assesses a building's environmental performance at the early stages of design. It addresses several topics, including airflow, natural daylight, thermal comfort, shadow analysis, sunlight exposure, and solar radiation.

The program's information processing process looks like this:
The user first generates a 3D model with Rhino/Grasshopper, then inputs details about the building form, orientation, material properties, and opening locations. At the same time, the application automatically loads climate data (EPW: Energy Plus Weather File) for the selected location. This comprises yearly values for temperature, humidity, solar altitude, radiation,

Feature	Details	Description or Sources
Wall material	Adobe and mud	Architectural articles or field surveys
Wall thickness	45 cm	Field surveys
Roof type	Vaulted and domed	Architectural articles
Roof thickness	20 cm	Field surveys
Openings features	1.2×1.5 meters (average)	Architectural drawings or measurements
Central courtyard size	250 m²	Architectural plans
Windcatcher dimensions	6 m	Field surveys
Building area	800 m²	Available drawings and documents
Window type	Double-glazed glass	Articles and field surveys
Door type	Wooden and traditional	Field surveys

Figure 10 : Architectural and Structural Features

Scenario	Parameter	Predicted Changes	Data Source
RCP4.5	Average temperature	Increase by 1-2°C	IPCC and CMIP6 reports
RCP8.5	Average temperature	Increase by 2-4°C	IPCC and CMIP6 reports
RCP4.5	Relative humidity	Decrease by 10%	IPCC
RCP8.5	Relative humidity	Decrease by 15%	IPCC
RCP4.5	Wind speed	Minor change	Local data and GCM
RCP8.5	Wind speed	Increase by 5%	Local data and GCM

Figure 9 : Climate Scenarios for Future Simulation

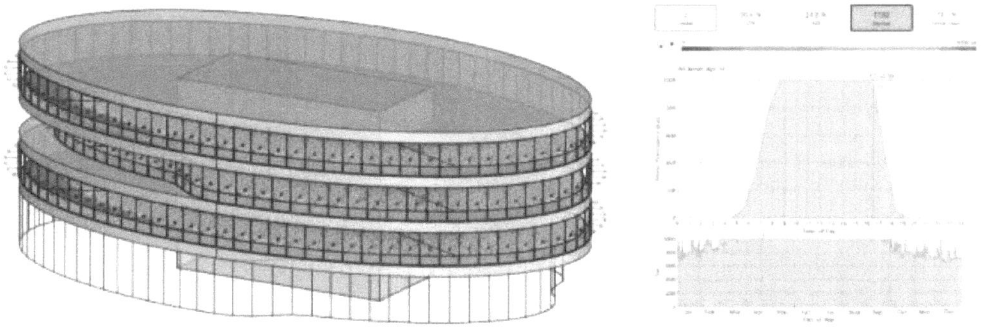

Figure 11 : Compatibility with Rhino, Grasshopper

wind direction and speed, and other environmental parameters. Climate Studio then uses a variety of simulation engines and algorithms to gather data. For instance: The Radiance engine uses a ray-tracing simulation technique to perform solar analysis. This determines how sunlight is reflected, absorbed, or transmitted over time as well as how it reaches interior spaces and building surfaces. Time-step spatial mapping Algorithms, which generate accurate data on annual sunlit hours, solar irradiation (kWh/m2), and shadow duration, are used in shadow analysis and sunlight exposure.

Thermal comfort analysis uses a multivariable thermal environment simulation algorithm that accounts for external conditions, window operability, ventilation rates, and internal heat gains. The user is given a visual representation of the analysis data that was produced. The program's compatibility with Rhino/Grasshopper, which enables the simultaneous creation of architectural forms and performance simulations, is one of its main advantages. Any design modifications are instantly reflected in performance analysis thanks to its real-time design feedback loop. The user can swiftly determine the best design strategies by modifying design parameters, like window size, orientation, and layout, in real time based on these results.

Conclusion

This study looked at how, in the face of shifting technological environments, modern architectural planning is developing around data-driven thinking. It examined the operational principles and methods of implementation of digital tools used for data collection, analysis, and interpretation, with a particular focus on the different types of data used in the early stages of design, specifically Pre-Design & Feasibility Planning and Strategic & Conceptual Planning. Every sophisticated analytical tool examined in this research is tailored for gathering specific kinds of data. Designers can produce precise and legitimate optimal design proposals based on the output these tools provide. Designers can quickly access the data that best suits their needs when they choose the right program based on the analysis's goal.

As a result, these tools make it possible to integrate non-architectural information like risk management, sustainability, economic viability, and user experience with architectural information. This suggests that modern architectural planning has evolved into a sophisticated decision-making process that is no longer dependent only on experience or intuition. As a result, this paper highlights how future architectural design must integrate data-driven tools and simulation technologies to create a more strategic and logical framework for decision-making.

Figure

Figure 1 : ArcMap : querying spatial data with relationships among geographic features
FIgure 2 : Overlay Function
Figure 3 : ArcCatalog : Framework for organizing large and diverse stores of GIS data.
Figure 4 : Defining the Different Cluster for Soil Modeling
Figure 5 Three Dimensional Models of the Soil and Pile with Meshing
Figure 6: : Incremental Shear Strain
Figure 7: Product's life cycle
Figure 8 : Buildings' life cycle stages according to EN standards
Figure 9: Climate Scenarios for Future Simulation
Figure 10: Architectural and Structural Features
Figure 11: Compatibility with Rhino, Grasshopper

Reference

Bob Booth and Andy Mitchell (2001) *'Getting Started with ArcGIS'*
GIS by ESRI, United States of America, pp.1-55
Available at:
https://webfacstaff.sas.upenn.edu/~dromano/classes/gis/files/Getting_Started_with_ArcGIS.pdf

Dhruba Lal Pradhan (2012),' *Development of P-Y Curves for Monopiles in Clay using Finite Element Model Plaxis 3D Foundation'*
Norwegian University of Science and Technology, Department of Civil and Transport Engineering, June 2012, pp.1-70
Available at:
https://ntnuopen.ntnu.no/ntnu-xmlui/bitstream/handle/11250/232178/566192_FULLTEXT01.pdf?sequence=1

Korpi, Yolanda (2024) *'How can calculating emissions in construction Industry with one click all enhance Biodiversity?'*
Häme University of Applied Sciences, Autumn 2024, pp.1-37
Available at:
https://www.theseus.fi/handle/10024/868248

Rezan Zarei (2024), *'Assessing the impact of climate change on building energy efficiency using Climate Studio software: a case study in hot and dry climate'*
Nov 2024, pp.1-16
Available at:
https://papers.ssrn.com/sol3/papers.cfm?abstract_id=5034857

Climate Studio official site : https://www.solemma.com/climatestudio

Planning in Building Science

Antonia A. Hochleitner[1], Lucia Sole[2], Hanna Weltken[3]
[1,2,3]Vienna University of Technology Wien
[1]e12223921@student.tuwien.ac.at [2]e12307576@student.tuwien.ac.at
[3]e12331435@student.tuwien.ac.at

The aim of the presented paper is trying to shape the concept of Building Science, an inherently interdisciplinary field which gains more and more prominence in the architectural context. The complexity in defining Building Science as a whole will be strictly connected to the variety of the challenges and questions that arise while designing. Building Science helps in studying and analyzing real-life problems throughout a scientific approach, thus gaining data and transferring information which enriches architectural design. The dialogue between science and architecture is contextualized in nowadays climatic crisis, shedding light on the importance of Life Cycle Assessment process and giving emphasis to the importance of inhabitants' wellbeing, which needs to be ensured during the whole life of a building and in city outdoor environments which are constantly changing. The urgency to quantify the use of nonrenewable resources in building constructions and to assess the environmental impact of the building is also a key part of national norms and standards and European guidelines. This has contributed to making Building Scientists key figures in heterogenic design teams and has encouraged the use of simulation and analysis tools also in early design stages.

Keywords: *Building Science, Evidence-Based Design, Performance Simulation,*

INTRODUCTION

The challenges facing contemporary architecture are more complex than ever. Climate change, resource scarcity and increasing demands for comfort and well-being are reshaping how we design and evaluate the built environment. In this context, the role of Building Science is gaining prominence. Not merely as a technical supplement to design, but as a foundational framework that connects aesthetic ambition with measurable performance.

THE ROLE OF BUILDING SCIENCE

Building Science is an interdisciplinary field that connects architecture, civil engineering, energy efficiency and sustainability. It bridges creative design with scientific analysis, helping designers to develop spaces which are also functional, comfortable and environmentally responsible. The increasingly more explored and documented correlation between well-being and environmental design, which is an implicit thought in design practices, sheds light on the importance of data-driven or evidence-based in architecture (Jackson, 2003). Building scientists play a primary role in these informed processes by understanding tools and methodologies to assess the project's performance from a systemic point of view, i.e. by considering the built environment as a complex system where indoor and outdoor conditions are correlated and influencing each other.

What Kind of Science is Building Science?

Assessing the indoor environmental quality of a building means both forecasting and analyzing: by forecasting, one can foresee the performance of a case-study at a specific point in time, while analyzing means to understand a current situation by gaining data from historical sources or existing case-studies. The two things are not separated, analysis is the foundation of forecasting, because by analyzing existing data, trends can emerge (Khan & Mir, 2021). Using existing data and forecasting at various points in time, makes it possible to predict the behavior of the building system in response to the outdoor environmental condition, and prediction is a key point of building science and it is crucial in creating informative frames to envision the outcome of some design decisions (Bomberg, 2012). The performance is evaluated through both user-related and cost-related indicators, where costs are both environmental and monetary. For such interest in both inhabitants and resources, building science is both a hard and a soft science: while sensing and simulations can be used to quantify measurable variables, such as temperature or humidity, and rely on mathematical models, to evaluate comfort conditions and occupants' satisfaction requires methodologies often applied in social sciences, such as interviews and surveys. Thus, building science is inherently an interdisciplinary field and its charm is in the ability of correlating multiple factors to evaluate the complexity of the built environment with rigorous approaches.

Finally, building science must be intended mostly as an applied science as it is related to practical goals, it takes theoretical knowledge from a variety of fields and applies it to evaluate real-world problems. The applied nature of building science in informing real design processes establishes a cyclical relationship between analysis and forecasting: when a design choice is guided by prediction, its implementation can be analyzed and evaluated, allowing for iterative refinement. This dynamic interplay has driven the emergence of new fields such as adaptive design and real-time system optimization, enabling buildings to sense and modify their behavior.

A Systematic Approach to Complex Design Challenges

A key challenge in architectural design is that problems are rarely isolated. Instead, they are part of a complex connection between parameters. When a single factor is changed, it often shows effects in several different areas. For example, placing a window affects daylight levels, thermal comfort, acoustics, the building's energy performance and user-related concerns, such as privacy.

Because of these interdependencies, designers must consider multiple variables simultaneously, which quickly leads to an overwhelming number of possible combinations. To manage this complexity, a systematic and evidence-based approach is essential.

This is where Building Science comes in. It offers tools, methods and frameworks to evaluate design decisions using measurable data and performance simulations. It supports an iterative design process, where individual solutions are assessed and compared. The goal is to optimize the overall building system within the constraints of technical, environmental and user-related goals (Pohl, 2011).

From Intuition to Evidence-Based Design

This not only improves design quality but also strengthens the ability to communicate and justify those decisions to stakeholders based on facts. Building Science is therefore often referred to as "Evidence-Based Design Support". Rather than relying solely on intuition, designers apply scientific knowledge and objective performance criteria to guide their decisions, e.g. the daylight factor, thermal comfort models or acoustic simulations.

One major advantage of Building Science is that it enables the quantification of subjective experiences that are traditionally difficult to measure, such as comfort. To achieve this, performance indicators have been developed to make these qualitative aspects comparable and objective.

A well-known example is the PMV (Predicted Mean Vote) and PPD (Predicted Percentage of Dissatisfied), which are used to quantify thermal comfort in indoor spaces. These standardized metrics allow designers to assess and compare environmental conditions numerically, helping them to make data-driven decisions during the design process. By using such indicators, comfort becomes a measurable, predictable part of performance-based design (Gilani, Khan and Pao, 2015).

Building Science as an Academic Discipline

Building Science is not just an abstract concept, but also an academic discipline with real-world application. The Technical University of Vienna (TU Wien) highlights the importance of Building Science in meeting today's architectural and environmental challenges by offering the master's program "Building Science and Environment". Students in this program are taught "how to apply scientific methods and state-of-the-art technologies in order to address the multi-faceted challenges posed by high performance and sustainable buildings" (TU Wien, n.d.).

This underlines the interdisciplinary nature of Building Science. The field encompasses a variety of specialized areas, each contributing to the creation of environmentally responsible, technically sound and human-centered architecture. Below are some key focus areas within the discipline (Kotharkar, Sarda and Deshpande, 2023):

- Sustainable Design and Green Building Practices
- Thermal Comfort
- Green Landscape and Green Site Planning
- Climate-Responsive Building Envelope Design
- Energy Efficiency in Building Services
- Net-Zero Energy and Carbon-Neutral Building Concepts
- Integration of Renewable Energy Systems
- Resilient Building Design for Extreme Weather Events

Life Cycle Assessment: A Key Tool for Sustainability

One of the most powerful tools is the Life Cycle Assessment (LCA). LCA is a method used to evaluate the environmental impact of a building across its entire lifespan: from raw material extraction, production, construction, operation, maintenance and eventual demolition or recycling. It quantifies energy use, greenhouse gas emissions and resource consumption, giving a comprehensive picture of a building's environmental footprint.

As such, LCA is an essential component of Building Science. It allows for the quantitative evaluation of environmental performance and helps designers make informed, data-driven decisions about materials, technologies and systems. LCA is supported by Building Science tools, like performance simulation.

Both LCA and Building Science emphasize the importance of early-stage integration. Decisions made in the early design phase have the greatest potential to optimize performance and minimize long-term environmental impact. Therefore, the application of Building Science has to be a fundamental part of the planning process (Kovacic, Reisinger and Honic, 2018).

IMPACT OF BUILDING SCIENCE ON THE PLANNING PROCESS

Building Science fundamentally changes the planning process by systematically and directly integrating the principles of physics, ecology and materials into the architectural and engineering design phases. In contrast to traditional planning methods, the planning process based on Building Science focuses on data-driven decisions. The use of advanced tools such as performance simulations, modelling and quantitative analysis enables planners to predict building performance accurately and optimize designs accordingly. These facilitate the evaluation and comparison of multiple design options at an early planning stage.

Interdisciplinary Collaboration in Early Planning Phases

Building science requires close interdisciplinary collaboration between architects, civil engineers, building physicists and sustainability experts from the very beginning of the planning process. This interdisciplinary approach is particularly relevant in the early planning phases, which are defined by standards such as the service phases (LPH) of the service model object planning - architecture (Leistungsmodell Objektplanung – Architektur) by the Federal Chamber of Civil Engineers (Bundeskammer der ZiviltechnikerInnen) in Austria (Lechner und Heck, 2014):

- LPH 1 Basic analysis: initial data collection, site analysis and determination of project boundary conditions.
- LPH 2 Preliminary design: Early conceptual development in which initial building physics findings, such as energy goals and sustainable strategies, are taken into account.
- LPH 3 Design planning (design planning/ system and integration planning): Detailed integration of building systems, performance simulations and optimization processes.
- LPH 4 Submission Planning: Preparation for official approvals and certifications (e.g. DGNB, ÖGNI), including detailed energy and building component certificates.

By integrating building science early on in these phases, planners can ensure that sustainable and efficient services are deeply embedded in the building concept, significantly reducing subsequent adjustments, cost overruns and performance deviations.

Norms and Regulations Influencing the Planning Process

The construction industry is heavily influenced by established standards and regulations that set clear performance and quality standards that must be adhered to during planning. This includes a hierarchical structure of standards ranging from international guidelines to local building regulations:

1. International Standards (ISO): Provide universally applicable methods and benchmarks for performance assessment.
2. European standards (EN): Ensure harmonization of performance metrics within Europe, which is critical for cross-border compliance and consistency.
3. National standards (e.g. ÖNORM in Austria): Specification of detailed national requirements, such as thermal performance (ÖNORM EN ISO 13370) or energy efficiency standards (ÖNORM H 5055).
4. Technical guidelines (e.g. the OIB guidelines in Austria): Provide practical guidance and specific compliance criteria, such as OIB Guideline 6 on energy conservation.
5. Local ordinances (e.g. Vienna Building Code): These deal with city-specific requirements

that tailor more comprehensive standards to local climatic and legal conditions.

Figure 1 provides a visual representation of the layered system of regulations, from international standards to local ordinances.

Compliance with these standards and regulations is simplified and rationalized by Building Science, as measurable performance criteria and standardized assessment tools provide clear evidence of conformity.

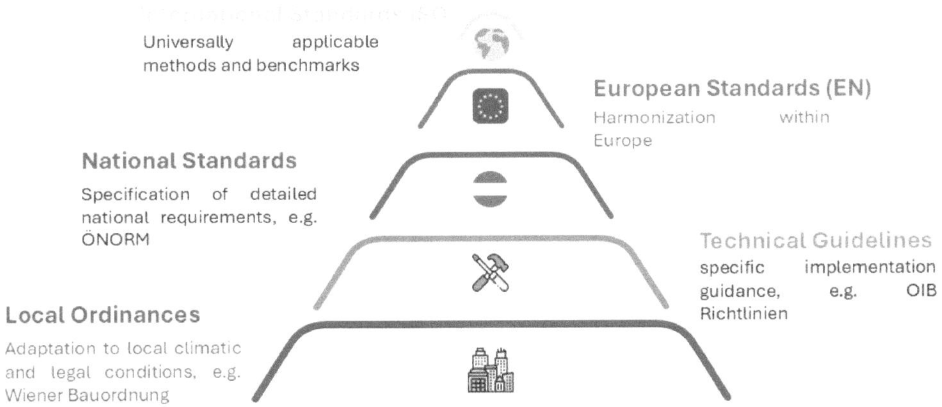

Figure 1: Hierarchy of Regulatory Standards

CONCLUSIONS

Building Science plays an important role in bridging the gap between architectural vision and measurable performance. As buildings are increasingly expected to respond to complex environmental, social and technical demands, new design processes need to be developed. By introducing evidence-based methods and simulation tools into early planning phases, Building Science enables designers to make informed, data-driven decisions that enhance sustainability, efficiency, and user comfort.

The interdisciplinary nature of the field encourages collaboration across architecture, engineering, and environmental sciences, fostering a holistic approach to design. Through the integration of standards, regulations, and quantitative assessment methods such as Life Cycle Assessment can be used.

REFERENCES

Jackson, L. E. (2003). The relationship of urban design to human health and condition. *Landscape and Urban Planning*, *64*(4), 191–200. https://doi.org/10.1016/S0169-2046(02)00230-X

Khan, A., & Mir, M. (2021). *Forecasting*.

Bomberg, M. (2012). Building science or building physics. *Frontiers of Architectural Research*, *1*, 421–423. https://doi.org/10.1016/j.foar.2012.10.003

Pohl, J. (2011) *Building Science: Concepts and Applications* [Online]. John Wiley & Sons. Available at: https://books.google.at/books?id=Ffh7O1F0KuoC&printsec=frontcover&hl=de&source=gbs_ge_summary_r&cad=0#v=onepage&q&f=false (Accessed: 24 March 2025).

Gilani, S.I.U.H., Khan, M.H. and Pao, W. (2015) 'Thermal Comfort Analysis of PMV Model Prediction in Air Conditioned and Naturally Ventilated Buildings', *Energy Procedia*, 78, pp. 859–864. Available at: https://doi.org/10.1016/j.egypro.2015.07.218 (Accessed: 24 March 2025).

TU Wien (n.d.) Building Science and Environment – Master's Programme [Online]. Available at: https://www.tuwien.at/en/studies/studies/master-programmes/architecture/building-science-and-environment (Accessed: 24 March 2025).

Kotharkar, R., Sarda, R. and Deshpande, R. (2023) 'Assessing the Integration of Building Science in Higher Education Curricula: Implications for Climate Change Adaptation in the Built Environment', in *CATE 2023: Comfort at the Extremes*. December 2023. Available at: https://doi.org/10.62744/CATE.45273.1188-161-171 (Accessed: 24 March 2025).

Kovacic, I., Reisinger, J. and Honic, M. (2018) 'Life Cycle Assessment of embodied and operational energy for a passive housing block in Austria', *Renewable and Sustainable Energy Reviews*, 82(Part 2), pp. 1774–1786. Available at: https://doi.org/10.1016/j.rser.2017.07.058 (Accessed: 24 March 2025).

Lechner, H. und Heck, D. (2014). *Leistungsmodell Objektplanung – Architektur [LM.OA]*. Graz: PMTools Software-Seminare-Verlag. Available at: http://www.pmtools.eu (Accessed: 24 March 2025). ISBN: 978-3-9503385-9-1

Types of Planning Tasks

Ângelo Miguel Pereira do Carmo[1], Besfort Ramadani[2], Ylleza Hoda[3]
[1,2,3]TU Wien
[1,2,3]{e11722580|e12402555|e12402551}@student.tuwien.ac.at

Architectural and urban design competitions are platforms to explore sustainable and interdisciplinary solutions for urban and architectural problems. This paper analyzes the planning tasks involved in three contemporary architectural competitions: Microhome 2025, House of the Future 2024/25, and the Dubai Urban Elements Challenge. A comparative taxonomy will be developed, characterizing planning tasks according to the planning stages, item types, life cycle phases and stakeholders' involvement. Planning complexity will be highlighted, interdisciplinary collaboration, stakeholder integration and sustainability strategies. Findings aim to support future competition future competition participants in aligning their work with current urban and architectural competitions.

Keywords: *Architectural competitions, sustainability, interdisciplinary collaboration, urban resilience, life cycle stage, and stakeholder engagement.*

INTRODUCTION

Architectural and urban design competitions encourage innovative and interdisciplinary solutions for issues such as climate change and urbanization. This facilitated the conceptualization and realization of architectural and urban interventions. The planning tasks required involve various project stages, participants, and goals. By systematically analyzing tasks, participants can design comprehensive and sensible proposals.

This paper looks into planning and execution stages in three distinct international competitions: Microhome 2025, House of the Future 2024/25, and Dubai Urban Elements Challenge. We will develop a taxonomy to categorize planning tasks, and make comparisons across the cases, focusing on planning stages, planned item, life cycle phases and stakeholder and disciplinary engagement.

The competitions were selected based on the scale, programmatic focus and location. The research involved understanding competitions briefs and guidelines. Planning tasks were highlighted and analyzed in four thematic areas. This approach enabled comparison while understanding the unique goals and constraints of each competition.

PLANNING STAGES

The planning process in the competitions that were analyzed is divided into different stages, each with specific goals and stakeholders. For example, in the Microhome competition participants must design self-sufficient and sustainable microhomes.

In a similar way, in the House of the Future, the goal is to design innovative solutions that integrate technology, sustainability and flexibility. The Dubai Urban Elements challenge focusses on city innovations aimed at improving the quality of the public spaces.

In the initial stage, architects, urban planners, and building scientists serve as the primary contributors, and organizers, the municipal bodies and the citizens play support roles.

All competitions give emphasis to different areas. When having a look at the House of the Future, it can be seemed that designers consider energy performance, climatic conditions, and community needs, and in the Microhome competition, it can be understood that participants focus on detailed floor planning and scalability of the microhome clusters. The Dubai Urban Elements competitors work with a broader urban context, including connectivity and social impact of their designs. Generally, there is collaboration between architects, environmental consultants and sustainability experts, and city planners, developers and community members act as stakeholders.

During the design development and the specialist planning phase, participants improve their concepts with architectural drawings and engineering details. For example, in the Microhome, entries add principles of energy and material efficiency, along with modular construction. In the House of the Future, competitors explore passive and active energy systems, circular construction methods, and adaptable layouts. At this stage, specialists, such as structural, mechanical and electrical engineers, works closely with the architectural team to ensure that the proposal is technically viable. In the Dubai competition, urban designers and infrastructure specialists contribute to detailed planning of the infrastructure, improving pedestrian sidewalks, landscaping and accessibility.

During the implementation stage, the primary concern is with construction, although not a central focus of design competitions, participants often propose strategies for construction schedules, methods, and quality control.

Finally, in the post-occupancy phase, long term performance is addressed, including future energy efficiency, retrofitting possibility, and user satisfaction. In both the Microhome, and House of the Future, competitions include future-proofing elements, allowing designs to be adapted and updated to meet evolving social and environmental needs.

Table 1 Taxonomy of planning tasks in Architectural Design Competitions.

Category	Descriptions
Planning Stages	Initiation, Design Development, Implementation, Post-Occupancy
Types of Planned Items	Micro-scale (Microhomes), Mid-scale (Co-housing), Urban-scale (Plazas, Parks, Networks)
Life Cycle Phases	Design & Construction, Use & Operation, Post-Use (Retrofitting, Monitoring, Adaptability)
Disciplines Involved	Architecture, Urban Design, Engineering, Building Science, Landscape, Digital Design
Stakeholders	Municipalities, Housing Agencies, Residents, Experts, Developers, Utility Providers
Scale and Focus	Microhome: Efficiency; House of the Future: Adaptability; Dubai: Connectivity & Public Space
Sustainability Strategies	Circular design, Passive systems, Prefabrication, Energy efficiency, Community integration
Interdisciplinary Collaboration	Designers, Engineers, Planners, Scientists collaborating across all competition stages

TYPES OF PLANNED ITEMS

Tasks are planned in various ways, depending on the type of item being designed, for example, in the Microhome competition, the planning goal is a Microhome, with a maximum floor area of 25 m^2. Designers also explore other ideas, such as modular construction, adaptability and living off-grid.

In the House of the Future, planning includes single-family houses, multi-family buildings and co-housing models. It is necessary to consider not only the individual houses, but also shared facilities, neighboring units, relationships within the community, and sustainable construction practices.

On the other hand, the Dubai Urban Elements Challenge focuses on the urban scale, considering interventions on parks, plazas, pedestrian networks, and waterfront development. Designers must consider social interaction, accessibility and resilience within the urban fabric.

Strategic planning and master planning are important for the Dubai competition, participants are asked to consider long-term use, infrastructure and connectivity strategies.

Altogether, the types of planned items highlight the scales and the challenges in architectural competitions.

LIFE CYCLE STAGE CONSIDERATIONS

Planning tasks can be separated according to their project's life cycle phase. The most publicly reviewed competitions were focused on new construction highlighting sustainable, technological, and modular design approaches. For example, the Microhome requested participants to design efficient microhomes that incorporate off-grid systems and circular construction principles.

Similarly, the House of the Future prioritizes adaptability in the long-term.

Future housing problems are often addressed in these planning tasks, retrofitting of existing spaces, and adaptability of the layout, and life cycle considerations are important during the post-occupancy phase. Tasks may involve monitoring energy use, enhancing building operations and ensuring occupant comfort.

The Dubai Urban Elements shows life cycling thinking, by setting usability and adaptability as its long-term goal.

In all these competitions, designers must implement life cycle considerations into their planning process. It is an essential step to create functional, sustainable and resilient solutions.

DISCIPLINES INVOLVED

Architecture and urban design competitions interact with many professional disciplines. Architects lead the design process, spatial integration, aesthetics, and functionality.

In the Microhome and House of the Future, building scientists contribute with expertise in energy performance, sustainable building materials, and indoor air quality. Engineers address technical aspects of building related to feasibility and building code compliance.

For the Dubai Urban Elements, planners and landscape architects focus on the urban context like pedestrian-friendly streets, biodiversity, and green infrastructure. The participation of digital designers, and material scientists, reflect technological innovation, prefabrication and modular construction design practices.

STAKEHOLDERS

Stakeholders' engagement is essential for architectural competitions. In initiatives such as the House of the Future and Microhome, the main stakeholder are municipalities, housing authorities and developers. They are generally responsible for naming the competition goals and serving as jury to evaluate the entries. Other entities like environmental agencies and utility providers can also be involved, providing expertise on ecological impacts, infrastructure capacity and energy performance.

When dealing with competitions that aim to improve community well-being and quality of life, the public is another important stakeholder. In the Dubai Urban Elements, participants are encouraged to consider how their designs can impact improve public spaces, promote social interaction and build resilience.

In the post-occupancy phase, residents and facility managers also become stakeholders, their feedback helps with assessing how well the built environment works overtime, by assessing usability, adaptability and long-term sustainability.

DISCUSSION

These competitions serve as platforms for exploring innovation in planning approaches, improving interdisciplinary collaboration and addressing global challenges such as climate change, urbanization and resources scarcity. This paper looks at planning tasks in three different competitions, using taxonomy and classifying these tasks according to planning stage, item type life cycle phase and stakeholder involvement.

These types of competitions encourage consideration of the full life cycle of buildings and urban spaces, rather than just the planning and building stage. Success is dependent on the active participation of the parties involved, namely the municipalities, developers and public.

The analysis and categorization of planning tasks within architecture and urban design competitions contribute to the broader discourse on sustainability, and resilient urban development. With the evolution of these competitions, they help redefining boundaries of design.

CONCLUSION

The Dubai Urban Elements challenge, House of the Future 2024/25 and Microhome 2025 all reflect the need to rethink how we design buildings and public spaces, considering environmental, social and technological changes. These competitions are different in scale but share a common goal: to push boundaries of conventional design through sustainability, adaptability and interdisciplinary collaboration.

By analyzing the planning tasks of each competition, this report shows how competitions can be more than just design exercises. They can test opportunities for complex ideas of real-world conditions. It can be by integrating local urban elements in Dubai, imagining the domestic life of future generations, or designing compact housing, each competition asks designers to consider long-term environmental impact, human needs, and collaboration processes that shape resilient architecture.

The value of these events is not just in the proposals, but in how they create a mindset of innovation, systems thinking, and urbanism. As global challenges increase, design competitions will help with finding solutions rooted in practicality, driven by creativity and informed by a collaborative multidisciplinary group.

REFERENCES

Microhome Kingspan 2025 Competition Brief: Buildner Architecture Competitions, in partnership with Kingspan. (2025). *Microhome Kingspan 2025: Competition Brief.* Available at: https://shorturl.at/8iOll Accessed: 28.03.2025

The House of Future 2024/2025 Competition Brief: Buildner, in partnership with Sheikh Zayed Housing Programme. (2024). *The House of Future 2024/2025: Competition Brief*. Available at: https://shorturl.at/pEji4 Accessed: 28.03.2025

Dubai Urban Elements Design Challenge Competition Brief: Buildner Architecture Competitions, Roads & Transport Authority (RTA) and Government of Dubai. (n.d.). *Dubai Urban Elements Challenge: Competition Brief.* Available at https://shorturl.at/aRPfl Accessed: 28.03.2025

ISO (2019). *ISO 15392:2019 – Sustainability in building construction – General principles.* Geneva: International Organization for Standardization

GAN's in Floor and City Block Planning

Insights in how AI can aid and automate planning processes

Paul Scheipner[1]
[1]*Technical University of Vienna*
[1]*paul.scheipner@student.tuwien.ac.at*

Designing both building interiors and urban blocks traditionally relies on manual drafting, iterative reviews, and exhaustive rule-writing-processes that are time-consuming, error-prone, and difficult to scale when complex regulations or local customs apply. In recent years, Generative Adversarial Networks (GANs) have emerged as a powerful, data-driven alternative for automating layout generation at both scales. By learning from thousands of existing examples rather than encoding every building code or zoning ordinance by hand, GANs can capture the implicit expertise embedded in historic floor plans (Goodfellow et al., 2014; Ali et al., 2025) as well as the cultural and legal nuances of city blocks (Fedorova et al., 2021; Xu et al., 2021). We conclude by identifying remaining challenges and promising paths toward fully automated, regulation-aware tools that architects and urban planners can trust and customize.

Keywords: *Generative Adversarial Networks, GAN, Floor & Block Planning, Gen Ai*

INTRODUCTION

Whether one is arranging rooms within a house or plotting parcels in a neighborhood, the core challenge is the same: satisfy intricate rules while crafting layouts that feel natural to end users. Traditional methods rely on architects' manual sketches, rule-based checks ("hallways must be X feet wide," "doors must swing inward"), and iterative revisions-a workflow that becomes exponentially harder as each new code or local convention (e.g., Amsterdam's bike lanes) is added (Smith & Jones, 2018; Fedorova et al., 2021; Xu et al., 2021).

To overcome these bottlenecks, researchers have turned to GANs - which learn implicit layout conventions from data instead of hand crafting every rule (Goodfellow et al., 2014). Early GAN's captured general building outlines but ignored room adjacencies and often produced awkward shapes (Radford et al., 2016; Uzun et al., 2020). Conditional image-to-image models then translated sketches or color-coded masks into walls, doors, and windows, greatly improving realism but sometimes yielding soft, blurry boundaries (Isola et al., 2017; Zheng & Huang, 2018; Sato et al., 2020). More recent graph-based GANs (e.g., House-GAN) accept room-adjacency or parcel-graph inputs, ensuring that generated layouts respect functional relationships and local constraints (Nauata et al., 2020).

GANS'S IN FLOORPLANING

Traditional floor-plan design relies on architects' manual drafting and iterative revisions-a process that takes a long time, is prone to mistakes, and becomes especially difficult when complicated building codes and regulations come into play. Early attempts to automate this work used sets of explicit rules-say, "hallways must be at least X feet wide" or "doors must swing in a certain way"-but adding each new requirement made the system exponentially more complex, and it quickly became impractical for real-world projects. In response, researchers have turned to Generative Adversarial Networks, or GANs, which learn from examples instead of relying on hand-written rules. By studying large collections of existing floor plans, GANs can pick up on the unwritten know-how of expert designers and produce layouts that feel familiar without explicit programming (Goodfellow et al., 2014).

Automating floor-plan creation still has its unique challenges. In the past, rule- or agent-based systems were popular, but each additional regulation-such as those for hospitals with strict safety rules-multiplied the system's complexity by a large factor, making it hard to maintain or extend (Smith & Jones, 2018). By contrast, a data-driven GAN model can absorb these complexities simply by including more examples during training. Real-world plans embed subtle conventions-like which rooms tend to sit next to each other, common paths people walk, and minimum room sizes-so a GAN trained on thousands of plans can learn these patterns naturally (Ali et al., 2025).

The first GANs applied to floor plans were based on Deep Convolutional GANs (DCGANs). These models take a random "seed" and try to turn it into a plausible building outline. While they could capture the general look of buildings in the training set, they often produced rooms in odd shapes and ignored important relationships like how bedrooms connect to bathrooms (Radford et al., 2016; Uzun et al., 2020). In other words, DCGANs showed promise but couldn't yet replace careful human planning.

Next came conditional image-to-image GANs like Pix2Pix, which treat floor-plan creation as a translation task: you give the model a simple sketch or color-coded zones (for example, blue for bathrooms, yellow for kitchens) and it fills in the walls, doors, and windows. This approach produces much more detailed and usable layouts (Isola et al., 2017). However, because these models typically use basic "pixel-by-pixel" comparison rules during training, their outputs can look a bit soft around the edges-walls may appear blurry, and fine details like window frames can be smudged (Zheng & Huang, 2018; Sato et al., 2020).

To make sure rooms appear in the right places relative to one another, researchers developed graph-based GANs such as House-GAN. Instead of pixels, these models take a simplified "bubble diagram" showing which rooms should connect. They then produce floor plans that respect those connections, and studies have shown people find these layouts more realistic and varied compared to simpler GANs (Nauata et al., 2020). The catch is that the results are still flat images, which makes it hard to import them directly into architectural software that prefers clean lines and editable objects.

Another practical effort, NVIDIA's ArchiGAN stack, breaks the problem into three steps: first it draws the overall building shape based on a property map, then it divides that shape into rooms with doors and windows, and finally it suggests furniture layouts (Chaillou, 2019). This step-by-step pipeline lets architects review and tweak each stage, combining fast AI suggestions with human expertise. While the demos look impressive, there isn't yet hard data on how often

architects accept these suggestions or how much time they save.
More recently, people have looked closely at the "rules" used to teach GANs. Instead of simple pixel-difference measures, they've tried a more forgiving formula called the Huber loss, which balances emphasis between small and large errors. On a large test set of sample plans, GANs trained with Huber loss produced crisper walls, cleaner room boundaries, and more reliable results than those trained with conventional methods (Ali et al., 2025).

Together, these developments show a clear path forward: moving from rigid, rule-based tools to flexible, data-driven methods that learn directly from real plans, enforce room relationships, and offer human-in-the-loop control. The next big step will be turning these flat images into editable, vector-based layouts that plug straight into architects' software. Promising ideas include hybrid models that blend graph reasoning with image translation, new generative techniques that output vector graphics directly, and interactive interfaces where designers can guide the AI in real time. By bringing these advances together, GANs in floorplanning may soon provide architects with fully automated, regulation-aware tools they can trust and customize.

GAN'S IN CITY BLOCK PLANNING

Even though planning a city block operates at a whole different meta level than planing individual floor layouts, the underlying challenge of adhering to (legal) regulations and meeting inhabitants' needs is quite similar (Goodfellow et al., 2014; Fedorova et al., 2021). Consequently, just as floorplan automation shifted from rule-sets/agents to data-driven methods, researchers have begun exploring GAN's to learn city-block patterns directly from existing urban patterns and data. This – at least for a large part - sidesteps the combinatorial explosion of hand-coded constraints and (human)effort(Goodfellow et al., 2014).

Moreover, city blocks introduce an extra dimension of complexity: local culture. For instance, designing a block in Amsterdam might need to consider accommodating bicycle traffic, while city block in a US city might be more considerate to cars (Fedorova et al., 2021; Xu et al., 2021). As a result, any automated system must not only respect building-height limits, setback rules, and land-use ratios, but also must be able to consider these implicit 'rules'.

Early work demonstrated that even a relatively simple GAN can internalize a city's morphological rules. In one study, Fedorova et al. (2021) trained a GAN on OpenStreetMap footprints from Milan, Tallinn, and Bengaluru, and showed that, without any explicit programming, the model learned block-shape distributions, street-alignment tendencies, and plot-size variations unique to each city. Consequently, the generated block layouts felt contextually appropriate, whether in the narrow lanes of Tallinn's old town or the winding streets of Bengaluru (Fedorova et al., 2021).

Building on this, Xu et al. (2021) introduced BlockPlanner, which represents each parcel and its neighboring lots as nodes in a two-tier graph. By combining a lightweight graph-VAE with GAN components, BlockPlanner produces full 3D block proposals that honor both plot boundaries and adjacency relationships, while also suggesting building-height variations that match real zoning codes. In comparative evaluations, this graph approach resulted in a more realistic land-use distributions and smoother transitions between neighboring plots than pixel-only methods (Xu et al., 2021).

In yet a completely different approach, Text2City (Qin et al., 2024) reframes block planning as a text-conditioned diffusion problem. By feeding in descriptions like "mixed-use waterfront district with pedestrian promenades," alongside surrounding map context, the model regenerates coherent road networks and building footprints that blend seamlessly with existing urban elements. This one-stage approach not only accelerates block prototyping but also allows non-technical people to perform urban design through natural language. As most city planners are probably not programmers, this novel approach makes a lot of sense for ease of use and acessablity (Qin et al., 2024).

Meanwhile, Gan et al. (2024) proposed UDGAN, which marries style-learning GANs with multi-objective optimization. By capturing a planner's aesthetic preferences - such as courtyard proportions or facades - UDGAN can generate multiple block schemes and then rank them according to predefines metrics. These intuitive metrices and KPI's are something like solar access and walkability. As a result, designers gain a palette of viable options that align with both regulatory requirements and personal design intentions (Gan et al., 2024).

Finally, UrbanGenoGAN (Cheng et al., 2023) demonstrates how GANs can integrate with evolutionary algorithms and GIS-based spatial analysis. A GAN proposes initial block layouts, and secondly a genetic algorithm refines them for connectivity and green-space distribution, and additionally, GIS tools verify flood-risk and transit accessibility. In case-study comparisons, UrbanGenoGAN outperformed conventional manual and optimization-only workflows on both quantitative indicators and planner satisfaction scores (Cheng et al., 2023).

Together, these advances trace a clear progression: from early GANs that learned generic block shapes to sophisticated, context-aware frameworks that incorporate graphs, text prompts, style preferences, and spatial analytics. Although challenges remain-particularly in exporting raster outputs to editable vector formats-promising directions include hybrid graph-pixel architectures, diffusion models with native SVG outputs, and interactive interfaces that let planners iteratively steer the generative process.

CONCLUSION

The unrelated evolution of GAN-based methods for floor-plan and city-block generation has a similar trajectory: from early, unconstrained models to conditional, graph-aware, text-guided, and multi-stage systems that embed regulations, culture, and human expertise directly into the learning process. Despite already impressive gains in realism, coherence, and user control, a massive gap remains in converting rasterized outputs into editable formats that integrate seamlessly with architects' CAD tools or urban planners' GIS platforms.

Promisingly moving forward, the synthesis of graph and pixel representations, the rise of vector-friendly diffusion models, and the design of interactive interfaces promise to bridge this very problem. By uniting these advances, the next generation of GANs might become truly automated, regulation - and culture-aware layout tools that enable designers at all scales to create safe, efficient, and contextually resonant environments without requiring the operator to be a computer scientist. In the ideal outcome, these GANs and systems will enable architects and planners-removing redundant and repetitive work-rather than replacing them themselves.

Parametric Design vs. Algorithmic Planning

Exploring Geometry Generation, Problem-Solving, and the Futaure of Computational Design

[1]*Raphael Auffarth*, [2]*Dorina Jakupi*
[1,2]*Technical University of Vienna*
[1,2]*{e11702811|e11944045}@student.tuwien.ac.at*

Parametric design and algorithmic planning are different methods in computational design but, they are linked. While algorithmic planning uses rule-based logic to create solutions, parametric design utilizes adjustable parameters to develop adaptable models. This study investigates their variances, uses, and overlaps in the context of architectural design. We investigate how tools such as Grasshopper, Blender, and Dynamo influence creativity and problem-solving.

Keywords: *parametric design, algorithmic design, algorithmic thinking.*

INTRODUCTION

Parametric and Algorithmic planning often get mixed up with each other. However, they are not the same, but they have a lot in common. Clarifying this distinction is essential for practitioners and educators. Parametric methods translate design intent into variable-driven geometry, while algorithmic planning prescribes solution logic through explicit rules. Mastering both approaches empowers architects to tailor computational strategies to specific project goals and constraints. Parametric design uses, as the name already says, parameters that change the form or the function of a certain object. (Assasi 2019) These parameters may have stand-alone effects on the object or can infect each other in some way. If the second case occurs you could speak of an algorithmic use of parameters. Algorithmic planning on the other hand deals with solving a problem with high complexity by using pre-defined rules to output infinite possible solutions. From this, it can be concluded that algorithmic planning can get along by itself while parametric design needs some sort of algorithmic thinking behind it. (Vazquez n.d.)

DEFINING KEY CONCEPTS

By allowing designers to create complex and adaptable structures, computational design has brought about a revolution in architecture and engineering. In this field, parametric design and algorithmic planning have developed as two essential approaches, with each offering different advantages for the design process. Parametric design provides flexibility via adjustable parameters, whereas algorithmic planning emphasizes problem-solving based on established rules. To use computational tools effectively, architects and designers must grasp these concepts. This segment investigates the definitions, and principles, of parametric design and algorithmic planning, laying the groundwork for their use in contemporary architectural practice. Academic literature often categorizes computational approaches as descriptive versus prescriptive. Descriptive methods map design intentions into geometric or data-driven relationships, while prescriptive methods specify procedural logic to generate solutions. Recognizing when to employ each mode ensures effective tool adoption across conceptual and detailed phases.

What is Parametric Design?

Parametric design is defined as a method where design elements are governed by parameters, enabling designers to create adaptable models that change in response to input values. This approach focuses on relationships among geometric or non-geometric variables, allowing for quick iteration and variation.

Parametric design goes further by integrating external data sources such as environmental sensors, user behavior metrics, or real-time structural feedback. Imagine a facade system that not only adapts its perforations based on historical sunlight exposure but also adjusts in real time to shifting weather conditions recorded by local stations. Designers can embed rules that alter geometry when indoor temperature or occupancy levels change, converting static models into living systems that respond dynamically to their surroundings. This level of responsiveness enhances building performance and supports sustainable design goals, offering architects powerful tools to meet evolving performance standards. The result is a responsive architecture that continuously optimizes for sustainability, cost efficiency, and occupant well-being.

What is Algorithmic Planning?

Algorithmic planning is a method in architectural design where the design is developed using explicit, step-by-step computational instructions known as algorithms. These algorithms describe the logic and sequence of operations required to generate the design, ensuring that every aspect of the final model can be directly traced back to a specific part of the algorithm.

This approach gives the designer full control over the design process by encoding decisions and relationships into a structured computational framework. It enables the creation of complex and precise architectural forms, where the logic of the design is transparent and reproducible. By relying on clearly defined rules, algorithmic planning supports accuracy, repeatability, and a high level of customization in the development of architectural models.

Algorithmic planning encodes design goals, constraints and material properties into a clear sequence of computational steps, so every change in a parameter automatically updates the model. It can draw on simulation or external data such as daylight studies or structural loads to refine geometry in real time. This method makes sure the designs are precise, reproducible and optimised for performance while allowing rapid exploration of alternatives.

HISTORICAL CONTEXT AND DEVELOPMENT

The roots of parametric design can be traced back to the introduction of computer-aided design (CAD) systems in the 1960s and 1970s. The early systems allowed designers and engineers to model objects with variable inputs, introducing the idea of form control by parameters for the first time. Although the technology was rudimentary compared to the present day, it established the broad premise that geometry could be defined not just by hand but also by logical and numerical methods.

In the mid-2000s, parametric design tools entered mainstream architectural practice. Platforms like Grasshopper for Rhino and Autodesk's Dynamo

made visual scripting accessible to architects without deep programming skills. Firms began using these tools for façade systems, structural grids and bespoke components. Workshops and online forums helped spread knowledge and build a community around shared scripts and sample files.
Today, parametric methods are tightly integrated with performance analysis, building information modelling and digital fabrication. Real time feedback on energy use, structural behaviour and material cost allows teams to optimise designs early in the process. Cloud based platforms and collaboration plugins enable multiple stakeholders to contribute to a single parametric model. This convergence of computation and data continues to push the boundaries of what architects can design and build.
In the engineering industry, CAD software enabled parameter-based modeling where parameters like dimensions and constraints were flexible to provide updated outputs. These principles became the foundation of what eventually would be further expanded as parametric design.

The Ecosystem of Parametric Tools

The range of programs that deal with parametric design in architecture is huge, the most popular ones being Grasshopper for Rhino and Blender, who have very different concepts of how they deal with parametricism. Grasshopper for Rhino uses a node-based interface where you link components to drive geometry and data. Its tight integration with Rhino's NURBS engine means every adjustment updates the model in real time, and community plug-ins such as Kangaroo and Weaverbird add simulation, analysis and fabrication tools.
Blender's Geometry Nodes live inside a complete 3D suite, so the same procedural networks can feed into modelling, rendering and animation. Python scripting and free add-ons like Sverchok add custom tools and batch workflows, making Blender a cost-free, all-in-one environment for computational design through to final presentation.

KEY TOOLS FOR PARAMETRIC DESIGN.

Grasshopper is the most used software in architecture when it comes to parametric design, and not without reason. It is built as a visual programming tool in which you can choose components and drag them onto your canvas to connect to one another. By doing so one can define rules and relationships between elements, making designs more flexible and dynamic. The big advantage of designing Grasshopper in Rhino is that the original model done in Rhino will not be changed but can be adjusted in Grasshopper by a single number slider for instance. Another advantage of Grasshopper is its SDK (Software Development Kit) which allows developers to create plugins to extend the functionality of Grasshopper. One of those is Kangaroo which works with physics components to simulate physical forces such as wind or gravity.

Figure 1: Gaudis hanging chain model

With tools like those a physical design model like the one of Antoni Gaudi made for the Sagrada Familia (*Figure 1*) in Barcelona, Spain can be reproduced digitally (*Figure 2*).

Figure 2: virtual hanging model in GH

Blender, on the other hand, has become a solid tool for parametric design, especially with Geometry Nodes that allow procedural modeling via a node-based workflow. Like Grasshopper, this allows for dynamic and flexible designs that can be modified in real-time. Blender is highly versatile since it combines modeling, rendering, and animation on a single platform, unlike Rhino and Grasshopper. Like Grasshopper's SDK for addon development, it gives support for Python scripting to assist automation and customization. Blender's parametric capabilities are further improved by add-ons like Sverchok, which offers a workflow more similar to that of Grasshopper. As a result, Blender is an attractive open-source option for architects and designers who are looking into computational design.

CONCLUSION

As computational design continues to shape the future of architecture, understanding the nuanced differences between parametric design and algorithmic planning becomes increasingly vital. While both approaches rely on rule-based logic and data-driven workflows, they offer distinct methods of engaging with complexity. Parametric design thrives on adaptability, offering a flexible framework where variables can be adjusted to produce dynamic models. In contrast, algorithmic planning emphasizes precision through structured logic, enabling designers to generate outcomes with a high degree of control and reproducibility.

The integration of tools like Grasshopper, Blender, and Dynamo has made these methods more accessible and powerful, enabling users to leverage complex computational strategies. As the boundaries between disciplines blur, and the demands of performance, sustainability, and collaboration increase, a hybrid approach that draws from both parametric and algorithmic thinking may offer the most promise.

Embracing both paradigms equips architects and designers with a richer, more versatile toolkit—allowing them to navigate from conceptual exploration to detailed realization with agility and foresight. The future of design will not belong to one method alone, but to those who can skillfully combine the strengths of many.

REFERENCES

Assasi, Reza. 2019. "Parametric Design, A Historical and Theoretical Overview," 17–18.

Vazquez, Elena. n.d. "Teaching Parametric Design: Fostering Algorithmic Thinking through Incomplete Recipes." Accessed March 30, 2025. https://doi.org/10.1108/OHI-06-2023-0135.

Gaudí, Antoni. *Sagrada Familia Church*. 1882. *JSTOR*, https://jstor.org/stable/community.10811285 . Accessed 9 June 2025.

Algorithms for Optimization

Pros and cons of different types of optimization algorithms

Lucija Zupa[1], Sama Feiziazar[2], Bükre Öksüz[3]
[1,2,3]Vienna University of Technology
[1]{e12402346|e12333478|e12223424}@student.tuwien.ac.at

Today, designing low-energy buildings has become more important than ever. However, achieving this goal requires navigating between several competing priorities—such as reducing energy use, managing costs, maintaining user comfort, and minimizing environmental impact. Because it's often difficult to meet all these criteria at once, designers increasingly rely on performance simulation tools to support decision-making and optimization. This paper examines six commonly used optimization methods—linear, nonlinear, combinatorial, integer programming, dynamic programming, and metaheuristic algorithms—through the lens of architectural design. Each method is evaluated based on criteria such as stability, speed, data coverage, accuracy, and robustness. The aim is to better understand the strengths and limitations of these approaches and to reflect on which ones are more suitable in different design contexts.

Keywords: *optimization algorithms, architectural design, simulation-based design, performance criteria, design decision-making, design evaluation*

INTRODUCTION

When it comes to optimization algorithms, they are used to find the best possible solution to a problem by minimizing or maximizing an objective function. These algorithms serve as tools for solving complex problems, especially when multiple variables need to be considered simultaneously. By working with a defined set of parameters, optimization algorithms explore the design space to identify the most efficient solution. What differentiates them from traditional design methods is their reliance on models and data generated through simulations and experiments, allowing for a more informed and data-driven decision-making process.

Thanks to their ability to evaluate thousands of alternatives efficiently, optimization algorithms are widely applied across various disciplines, including architecture. In general, algorithmic optimization methods can be categorized into several types such as search procedures, loss function-based approaches, and convex programming, each of which takes a different approach and includes numerous subcategories.

This paper focuses on the application and impact of optimization in the field of building science, which has gained more attention over the past decade. One reason for its delayed adoption in design practice was the lack of accessible digital tools that could translate the complexity of architectural intentions into measurable outputs. However, the building sector today accounts for approximately 40% of global energy use and nearly 30% of CO_2 emissions, pushing the industry to adopt more efficient design methods (IEA, 2023). Facing challenges such as thermal comfort, lighting, acoustics, and material usage, architects increasingly turn to simulation-based design tools (Evins, 2013). The growth of parametric design platforms has made it easier to apply optimization techniques, from solving basic form-finding tasks to addressing more complex systems including environmental performance, automation, and even human-centered factors.

This paper is structured around six key types of optimization methods: linear, nonlinear, combinatorial, integer programming, dynamic programming, and metaheuristic algorithms.

METHODOLOGY

This study investigates six optimization techniques widely used in simulation-based architectural design. The selection of these methods was based on a literature review of academic papers, industry usage, and qualitative performance criteria such as stability, speed, coverage, and accuracy. For each method, both technical characteristics and practical implications in the context of architectural workflows were analyzed. The focus was on identifying key features and differentiating qualities of each method, including both their strengths and limitations. This helped to outline the most relevant performance aspects for architectural applications.

In the next step, six representative optimization techniques were selected for detailed analysis: linear optimization, nonlinear optimization, combinatorial optimization, integer programming, dynamic programming, and metaheuristic algorithms. These methods were chosen based on how well they reflect the variety of problems encountered in simulation-based design. Each technique was then assessed according to specific performance indicators such as stability, computational speed, data coverage, validity, accuracy, and robustness. These criteria allow for a more structured and comparative evaluation of the strengths and weaknesses of each optimization approach in architectural contexts.

PROS AND CONS OF OPTIMIZATION METHODS

In architectural design, selecting an appropriate optimization method depends on factors such as problem structure, data type, computational cost, and desired level of accuracy. This section introduces the main algorithm categories commonly applied in design optimization and discusses their key strengths and limitations.

LINEAR OPTIMIZATION

Linear programming is one of the most fundamental optimization methods, it is mostly used for maximizing or minimizing a linear objective function subject to a set of linear equality and inequality constraints. For the architectural design. It is typically applied in early-stage of planning, construction cost estimation, and material optimization, where relationships between variables—such as cost, area, or volume—can be represented in linear terms. Its deterministic nature guarantees the global optimum for problems that fit the linear methods. This method has been widely studied in engineering and operations research, and although its application in architecture is somewhat limited, it serves as a baseline method due to its computational efficiency and ease of implementation (Chellai, 2025).

Pros.

Linear optimization offers fast and exact solutions, especially when dealing with resource allocation or simple constraint-based problems. It is supported by a wide range of solvers and optimization tools, making it easy to integrate into the existing design processes. Its low computational cost and clarity of results make it an ideal method for feasibility studies and simplified parametric analyses (Chellai, 2025).

Cons.
Despite its strengths, linear optimization is limited in architectural practice because most real-world design problems involve non-linear relationships. Factors like structural deformation, daylight performance, thermal comfort, and acoustics rarely exhibit purely linear behavior. As a result, linear methods often oversimplify complex systems and fail to capture the richness and variability of architectural design processes (Liu et al., 2024; Chernova et al., 2024).

NONLINEAR OPTIMIZATION

Nonlinear optimization refers to the process of optimizing an objective function that includes one or more nonlinear relationships between variables or constraints. This method is particularly suitable for architectural design processes involving complex systems, such as energy modeling, structural simulations, daylight analysis, or airflow simulations, where the performance metrics depend on nonlinear physical laws. Unlike linear programming, nonlinear optimization does not guarantee a global optimum and often requires iterative algorithms and advanced solvers to converge to acceptable solutions (Liu et al., 2024).

Pros.
Nonlinear optimization provides the flexibility needed to accurately represent complex architectural performance problems. It allows for a more realistic modeling of design scenarios and enables the integration of simulation data, resulting in more informed design decisions. Its capacity to handle irregular and nonconvex problems makes it particularly relevant in sustainability-focused or high-performance design (Liu et al., 2024).

Cons.
The main drawback of nonlinear optimization is its computational intensity and the risk of converging to local, suboptimal solutions. The performance of the algorithm is often highly dependent on the initial conditions and solver settings. Moreover, it typically requires greater technical knowledge to implement and interpret, which may limit its accessibility in early-stage architectural design (Chernova et al., 2024).

COMBINATORIAL OPTIMIZATION

Combinatorial optimization deals with problems where the goal is to select the best solution from a finite but often exponentially large set of discrete possibilities. In architectural design, these problems typically arise in spatial layout planning, circulation optimization, modular configurations, and structural member arrangements. Each possible configuration corresponds to a different combination of design elements—such as room placements, wall connections, or façade modules—which must be evaluated based on a set of performance criteria. Traditional exact methods are often inefficient due to the high dimensionality and complexity of such problems. As a result, many researchers and designers have adopted heuristic and metaheuristic techniques, or hybrid frameworks, to navigate this discrete design space more efficiently (Liao et al., 2025; Rashid & Zebari, 2025).

Pros.
Combinatorial optimization is particularly suited for architectural problems that involve discrete variables and logical constraints. It can be applied to solve highly practical challenges in layout design, modular system generation, and structural topology. Additionally, it supports multiple objective functions, enabling balanced solutions across conflicting design

goals such as aesthetics, efficiency, and constructability (Liao et al., 2025).

Cons.
Due to the exponential growth of the solution space, combinatorial optimization problems are generally NP-hard and require significant computational resources to solve optimally. For large-scale architectural projects, exact solutions become infeasible, and approximate methods may not guarantee optimality. Furthermore, these problems often need customized encoding and careful constraint handling, which increases the complexity of the modeling and solution process (Rashid & Zebari, 2025).

INTEGER PROGRAMMING

Integer Programming (IP) is an instance of mathematical programming where certain or all decision variables are constrained to integer values. In practice, IP generally refers to Integer Linear Programming—optimisation problems with linear objective function and linear constraints and variables restricted to integer (typically binary 0/1) values. IP is an instance of the broader category of mixed-integer programming, heavily applied in operations research. It has been applied to resolve architectural design problems that can be discretised or modularised, such as space layout planning on grids, component selection, or construction scheduling. For example, Xu and Li (2019) formulated residential floor plan generation as an IP problem, positioning rooms (as pre-specified shape templates on a grid) subject to adjacency constraints to produce optimised layouts. In these applications, design objectives like minimizing empty space or cost of construction are recast as a linear objective function, and geometric and topological constraints are cast as linear constraints.

Pros

One of the greatest advantages of Integer Programming (IP) is the possibility of delivering a globally optimal answer (within model constraints), if solved to completion. Compared to heuristic approaches, IP is an exact method—so that, given sufficient computing time, an IP solver will either come up with the definitive best solution or prove that there is no better one, as opposed to merely approximating one (Wikipedia, n.d.-a). This level of optimality is especially relevant when the goal of design is quantifiable and when an absolute "best" solution must be obtained, as in the instance of minimizing structural cost under compliance with all codes of regulation.

IP reliability is evident in comparative studies. For instance, in on-site construction layout planning, a mixed-integer programming model created more cost-effective solutions than a Genetic Algorithm (GA), and with easier introduction of additional layout constraints—a prerequisite that made the GA-based approach more time-intensive (Wong, Fung and Tam, 2010). In such integer spatial arrangement problems, where facilities or rooms and their locations may be modeled as integer variables, IP has consistently outperformed GA in solution quality.

Another significant strength of IP is its capability to address complex constraint sets precisely. Architectural design typically comprises several "hard" constraints—such as spatial adjacency restrictions, dimensional requirements, or cost constraints—that must be imposed rigorously. IP is especially suitable for such conditions, as additional constraints can generally be incorporated into the model without altering the optimisation strategy in any essential manner. In contrast, incorporating the same conditions into a metaheuristic solution often involves the need to use penalty functions or post-processing repair methods (Wong, Fung and Tam, 2010).

Cons.

The primary drawback of IP is its computational cost and modelling rigidity, especially for complicated architectural problems. IP problems are NP-complete in general, which means the worst-case solution time grows exponentially with problem size (Xu and Li, 2019). When the design issues are big (i.e. many rooms or discrete options), or the solution space gets explosive, IP solvers may become impractically slow or unable to locate the optimum in a reasonable time frame. It is this absence of scalability that restricts the use of IP to comparatively well-structured or modestly sized issues. For instance, though an IP model can optimally place a dozen rooms on a grid, its performance deteriorates drastically with the complexity of the grid or the number of rooms.

Early surveys of layout planning characterized the floor plan layout issue as a challenging combinatorial problem, noting that "as yet there appears to be no algorithm available which produces a global optimum" for large-scale problems (Gero, 1975). This helps highlight how exact methods like IP, while theoretically appealing, can be overwhelmed by real-world design complexity. Practitioners are thus often compelled toward simplifications or coarse grid representations in order to maintain solvability—at the cost of design resolution.

Another limitation is that IP requires a precise mathematical formulation of the design problem that might be impossible to achieve for qualitative or subjective criteria. Objective functions such as an "aesthetically pleasing composition" or advanced environmental comfort objectives tend to be hard to quantify. Even when quantifiable, the relationships can be very non-linear or not susceptible to being cast as linear constraints. Wortmann and Nannicini (2017) observe that some elements of the optimisation problem (e.g., decision variables, objectives, constraints) may be difficult to detect or formulate, and even if detectable, might not be expressible algebraically.

DYNAMIC PROGRAMING

Dynamic Programming (DP) is an algorithmic strategy that solves problems by decomposing them into overlapping sub-problems, calculating each of the sub-problems only once, and storing their results to be reused. Classic DP methods, such as those based on the Bellman equation, use recursion or iterative table-filling to fully calculate all states of a problem, but in a manner that saves redundant computation (Wikipedia, n.d.-b). DP actually replaces brute-force enumeration with a more intelligent plan that takes advantage of optimal substructure—where the optimal solution for a problem can be formed using optimal solutions of its subpieces.

Some common uses of DP are standard problems such as shortest path finding (e.g. Dijkstra's algorithm), the knapsack problem, and sequence alignment. DP, however, in architectural design has not gained as much popularity as Integer Programming (IP) or metaheuristic algorithms have. DP has still been applied to some types of design problems that inherently appear in a staged or sequential decision-making process.

Past research during the 1970s and 1980s demonstrated DP to be effective in such applications as site planning, circulation route optimisation, and layouts of modular housing (Gero, 1975). For example, DP could be employed to determine the best order of positioning buildings on a site—optimising each placement step based on the aggregate layout—or determining the most effective vertical pattern of circulation in a high-rise building by successively adding stops to an elevator.

Pros

DP's greatest strength lies in its ability to find optimum solutions computationally effectively—if a problem has the right structure. That is, if a design problem displays optimal substructure (i.e., a global optimum can be constructed from optimum solutions to sub-problems), then DP can guarantee the global optimum while maintaining computation costs significantly below brute-force techniques (Wikipedia, n.d.-b). Technically, DP generally converts exponential search complexity into polynomial time via reuse of intermediate results and application on similar sub-problems (Wikipedia, n.d.-b).

DP comes especially strong for certain planning problems in building design. As a sample, if there was a space planning problem for departments that have to be assigned onto floors within a building, then a DP solution would be capable of optimizing floor-by-floor the department placements process, and cache and reuse part-solutions to sub-sets of placements. If properly articulated, the technique ensures that the final assignment is globally optimal. Unlike heuristic methods, which might miss the optimum due to local optima or randomness, DP systematically attempts all combinations necessary to ensure optimality.

DP has also been appropriately described as a clever exhaustive search technique. As one definition states, "a dynamic programming algorithm will consider all possible solution paths and will select the best solution... a clever brute-force strategy" (Wikipedia, n.d.-a). Where sub-problems overlap in problems, this is even more so. For example, the notorious 0–1 knapsack problem—exponential time when solved by brute force—can be solved in pseudo-polynomial time with DP by employing sub-capacity memoisation (Wikipedia, n.d.-b).

In building design applications, certain activities such as step-by-step planning or modular construction have been discovered suitable to DP. Gero (1975) further stated that DP has been meaningfully applied in campus site planning, vertical circulation planning, house planning, and site development. All of these applications involve making naturally following decisions in sequence—e.g., zoning a site one step at a time or distributing functions among floors—so they are naturally most beneficial to the capabilities of DP.

Cons

Despite its benefits, Dynamic Programming (DP) possesses severe limitations that restrict its broader use in building design. Its most serious limitation is that not all design problems possess the required structure to make DP effective. Specifically, DP relies on optimal substructure and overlapping sub-problems—attributes that are primarily absent in the highly interconnected and multi-dimensional nature of building design problems (Wikipedia, n.d.-b).

Architectural designs usually contain elements that influence one another in complex, nonlinear modes, and it is difficult to decompose them into small, autonomous phases. If broken down to such an extent, DP is vulnerable to what is referred to as the "curse of dimensionality"—an exponential growth of state space with more parameters (Wikipedia, n.d.-b). For example, attempting to use DP to position n rooms in sequence in a building plan would entail maintaining and storing an astronomical amount of partial configurations, which is computationally unfeasible when n increases.

As noted by Gero (1975), while DP was effectively implemented for certain specific architectural problems in the past, e.g., site phasing and campus plan configuration, it is still not effective in dealing with more generalized, open-ended design issues. Indeed, the inability to scale DP to tackle the floor plan issue as a whole was one of the first known limitations—"no algorithm produces a global optimum" in large, unconstrained configuration cases, even though DP is theoretically exact.

A second significant drawback is that DP needs a sharply defined, step-by-step decision model, which is inherently in opposition to the nonlinear, iterative process of architectural design thinking. Design processes usually involve backtracking, co-evolving criteria, and aesthetic judgment—none of which translate well into the strictly staged, forward-only paradigm necessary for DP to work.

Even in such issues where DP may be applied, it typically solves one objective at a time. Multi-objective optimisation—say, cost, spatial quality, and energy performance—is conquered by either simplification into one weighted objective or repeated Pareto methods, both reducing DP's utility in real-world design contexts.

METAHEURISTIC ALGORITHMS

Metaheuristic algorithms are high-level search methodologies that traverse complex solution spaces. In contrast with exact optimisation methods, metaheuristics provide no assurance of a global optimum but rather seek adequate solutions through iterative improvement and probabilistic search. Prominent examples in architectural design include the Genetic Algorithm (GA), Simulated Annealing (SA), and Particle Swarm Optimisation (PSO), which take inspiration from natural processes of evolution, thermodynamics, and swarm intelligence respectively. They are particularly well-adapted to domains with a vast, non-linear, and difficult-to-model design space analytically—features that are shared by architectural optimisation problems (Wortmann and Nannicini, 2017).

Metaheuristics treat the design problem as a "black box," in the sense that they require only a function that evaluates the quality of each solution. This allows them to readily couple with performance simulations—such as energy, daylighting, or structural analysis—so they are popular in simulation-based design optimisation (Wortmann and Nannicini, 2017). Practically, plugins like Rhino/Grasshopper's Galapagos (GA) solver enable architects to optimise parametric models interactively, without needing to program (Wikipedia, n.d.-c). Moreover, advanced evolutionary approaches such as NSGA-II enable multi-objective optimisation by generating a Pareto front of trade-off solutions within a single run.

Pros

The biggest advantage of metaheuristics is that they are adaptable. They make minimal assumptions about the structure of the problem and can solve black-box, non-linear, or discontinuous objectives. That is, they can be used to apply with simulations whose representation is not closed form (Wortmann and Nannicini, 2017). For example, a GA can optimize building energy performance by connecting with an EnergyPlus simulation so that it can compute each design based on actual performance data.

Another major strength is their potential for multi-objective optimisation. Metaheuristics, especially evolutionary algorithms, work with populations of potential solutions and have an intrinsic mechanism to generate a variety of different alternatives. Thus, they can approximate Pareto-optimal sets—offering trade-offs among conflicting design goals such as cost, comfort, and aesthetics (Caldas and Norford, 1999). Their use in creative design applications is also adequately documented: GAs have been employed for optimizing facade geometry, window placement, and even for designing novel structural forms that satisfy both performance and aesthetic expectations (Caldas and Norford, 1999).

Metaheuristics are also highly scalable and workflow-independent. Because they are merely solution evaluators, they can be applied to any performance measures-generating model, ranging from finite element models to acoustic ray tracing. They also naturally support constraint management through penalty, repair operators, and user-specified encodings, allowing even advanced design rules to be managed.

Cons

The primary weakness of metaheuristics is that they lack any form of global optimality guarantee. Being founded on stochastic processes, the solutions they come up with tend to be approximations and may converge to local optima. This is undesirable in key or highly constrained design cases where optimality is a priority (Wortmann and Nannicini, 2017).

Another concern is computational cost. Metaheuristics require checking out hundreds or thousands of solutions, especially when applied to high-computational simulations. Brute-force, this might result in excessively long run times, and because of that, they are less efficient than exact methods for simple or well-formulated problems (Wortmann and Nannicini, 2017).

Metaheuristics are also sensitive to parameter tuning—e.g., population size, mutation rate, or cooling schedule)—affecting speed of convergence and quality. Poorly chosen parameters can lead to ineffective searches. In addition, some problems require problem-specific representations (e.g., chromosome encoding in GA), and poorly designed representations will cause the algorithm to produce infeasible or inefficient solutions.

Because of the stochastic nature of them, solutions may differ for each run. This adds uncertainty and can be why runs or post-processing may be required to be robust in the solution. If the application is safety-critical, then the absence of deterministic result can be decreased trust in the solution. Metaheuristics are also less effective if the evaluation landscape is misleading or noisy, i.e., if small geometric alterations trigger large simulation fluctuations. In such cases, hybridisation based on local search or surrogate modelling may be required to boost consistency.

PERFORMANCE CRITERIA

In order to compare the selected optimization methods in a structured and meaningful way, a set of evaluation criteria was used. These criteria were taken from existing literature on optimization in design disciplines (Evins, 2013; Wortmann, 2017) and refined based on the specific needs of simulation-based architectural workflows. The focus is not only on mathematical accuracy, but also on how each algorithm performs in practical design scenarios. The following seven criteria were used in this comparative evaluation.

Stability
The stability of data describes the ability for the algorithm to produce the same values over and over when repeating the same analysis. By being one of the basic characteristics for data system descriptions, often it is used to describe optimization on the market as being a reliable product.

Speed
Speed reflects how quickly the algorithm can process a range of design variations within a given timeframe. The runtime of the optimization is closely related to simulation time and indicates the efficiency with which the algorithm explores the design space. This becomes particularly important in complex tasks where time-saving is crucial for reaching optimal solutions.

Coverage
The coverage demonstrates ability to find the solution in workable space meaning it is used to reflect search capability.

Validity
The validity is used to expose the capability of the algorithm to find the solution. This refers to finding an algorithm with true optimum meaning finding the correct solution for the problem it's working on. it should be evaluated in conditions such as the context of the data source and finding personalized solutions.

Efficiency
This algorithm is used to minimize the usage mostly having time and space in mind but still to achieve optimal performances. it can be applied when some process is repeating.

Safety
Safety is necessary to control and reduce by taking them into account properly and having in mind human safety as well as data protection.

Accuracy
Accuracy refers to the precision of the algorithm in decision-making, prediction, and solution generation. It determines how close the output is to the actual or ideal result and plays a defining role in the overall success of the optimization process.

CONCLUSION
This paper analyzed six different optimization methods applied in architectural design, focusing on their performance in simulation-based workflows. Each method—ranging from exact solvers to heuristic techniques—offers distinct strengths depending on the type of problem addressed. While integer programming and linear methods are more suitable for structured, constraint-heavy tasks, metaheuristic and nonlinear approaches provide flexibility for complex, high-dimensional design spaces. Future work may explore hybrid optimization strategies that combine exactness with adaptability, aiming to better support sustainability-driven design processes.

REFERENCES
Caldas, L. and Norford, L. (1999). A genetic algorithm tool for design optimisation. In: ACADIA '99: Proceedings of the 19th Annual Conference of the Association for Computer-Aided Design in Architecture. Salt Lake City, UT, pp.260–271.

Chellai, F. (2025). Optimization Methods (Non-Linear Programming). University Ferhat Abbas. Available at: https://www.researchgate.net/publication/389739744 (Accessed: 1 April 2025).

Chernova, L. and Georgieva, S. (2024). Linear vs. nonlinear optimization in environmental modeling. Environmental Modelling and Software, 163, 105776. https://doi.org/10.1016/j.envsoft.2023.105776

Evins, R. (2013). A review of computational optimisation methods applied to sustainable building design. Renewable and Sustainable Energy Reviews, 22, pp.230–245. https://doi.org/10.1016/j.rser.2013.02.004

Gero, J.S. (1975). Architectural optimisation – a review. Engineering Optimization, 1(3), pp.189–199. https://doi.org/10.1080/03052157508960586

IEA (2023). Buildings – Tracking Clean Energy Progress. International Energy Agency. Available at: https://www.iea.org/reports/buildings (Accessed: 1 April 2025).

Liao, Z., Chen, J., Wang, D., Zhang, Z. and Wang, J. (2025). Neural combinatorial optimization via preference optimization. arXiv preprint. Available at: https://arxiv.org/abs/2503.07580 (Accessed: 1 April 2025).

Liu, S., Hua, X., Shan, L., Wang, D. and Sun, Y. (2024). An improved particle swarm optimization method for nonlinear optimization. International Journal of Intelligent Systems, Article ID 6628110. https://doi.org/10.1155/2024/6628110

Rashid, N.S. and Zebari, I.M.I. (2025). A comprehensive review of metaheuristic algorithms for combinatorial optimization problems. International Journal of Scientific World, 11(1), pp.83–92. Available at: https://www.sciencepubco.com/index.php/IJSW (Accessed: 1 April 2025).

Wong, C.K., Fung, I.W.H. and Tam, C.M. (2010). Comparison of using mixed-integer programming and genetic algorithms for construction site facility layout planning. Journal of Construction Engineering and Management, 136(10), pp.1116–1128. https://doi.org/10.1061/(ASCE)CO.1943-7862.0000214

Wortmann, T. (2017). Ten questions concerning optimization in architecture. Building and Environment, 125, pp.575–586. https://doi.org/10.1016/j.buildenv.2017.09.017

Wortmann, T. and Nannicini, G. (2017). Introduction to architectural design optimisation. In: A. Karakitsiou et al. (eds) City Networks: Planning for Health and Sustainability. Cham: Springer, pp.311–332. https://doi.org/10.1007/978-3-319-65338-9_18

Xu, J. and Li, B. (2019). Searching on residential architecture design based on integer programming. In: Proceedings of the 24th International Conference of the Association for Computer-Aided Architectural Design Research in Asia (CAADRIA 2019). Wellington, New Zealand, 15–18 Apr. 2019. CAADRIA, pp.263–270.

Data Driven Resilient Architecture

Three Pillars of Resilient Architecture and Exemplary Applications

Elisabeth Großberger[1], Isabelle Seifert[1]
[1]TU Wien
[1]e52006960@student.tuwien.ac.at [2]isabelle.seifert@ student.tuwien.ac.at

In today's context of increasing environmental challenges, resilient architecture designed to withstand and adapt to environmental stresses is increasingly important to future-proof our built environment. This paper introduces 3 important pillars of resilient architecture and explores the role of data-driven approaches to enhance the resilience of architectural design. By the integration of real-world data, smart systems and predictive simulation, buildings can be planned that anticipate and adapt to potential disturbances. The combination of data science and architecture is seen as a promising approach for sustainable and resilient urban development.

Keywords: *Data-Driven Design, Resilient Architecture, Building Industry.*

INTRODUCTION

As climate change, rapid urbanization, and resource scarcity increasingly challenge the built environment, the concept of resilient architecture has gained significant relevance. In architectural discourse, resilience refers to a building's ability to resist, absorb, adapt to, and recover from environmental stressors while continuing to provide essential functions. Distinct from sustainability, which emphasizes minimizing long-term environmental impact, resilience prioritizes immediate and adaptive responses to disruption (Bulman, 2024).

The following sections provide a brief historical background and explore the growing necessity for resilient architecture. Furthermore, the three fundamental pillars of resilience, durability, adaptability, and sustainability are outlined. Finally, selected applications of data-driven and algorithmic design are presented to illustrate how emerging digital methods can support resilience in architectural practice.

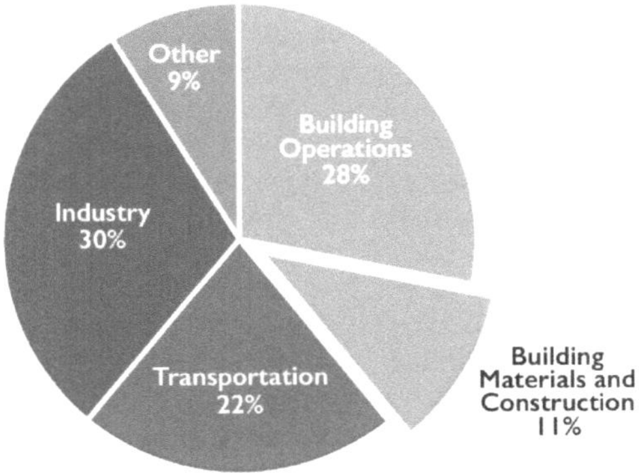

Figure 1
Global CO2 emissions by Sector (Architecture 2030)

HISTORY AND BACKGROUND

The AEC sector, responsible for nearly 39% of global CO_2 emissions (Fig. 1), is a major contributor to climate change.

Therefore, governments and international organizations emphasize the importance of reducing this number by creating more resilient buildings and constructions (Architecture 2030, n.d.). In alignment with the Sustainable Development Goals (SDGs), the United Nations Department of Economic and Social Affairs has established Goal 11, which focuses on making cities and human settlements inclusive, safe, resilient, and sustainable (United Nations, n.d.).

Resilience, as first defined by the National Academy of sciences, is "the ability to prepare and plan for, absorb, recover from, and more successfully adapt to adverse events." (Connelly et al., 2017).

Historically, resilience in architecture was embodied in vernacular forms. For example, raised, stilt structures were common in flood-prone areas, wind-resistant forms in tropical regions, and thick-walled constructions in desert climates served for passive thermal regulation. In contrast, contemporary resilience is increasingly shaped by technological innovation, advanced materials, and data-driven design strategies.

THE 3 PILLARS OF RESILIENT ARCHITECTURE

Climate-resilient architecture integrates three core principles: durability, adaptability, and sustainability. These pillars ensure that buildings remain functional, efficient, and environmentally responsible throughout their lifespan.

Durability

Durability refers to the capacity of architectural systems to maintain structural integrity and functionality over time under physical, environmental, and operational stress. It underpins resilience by minimizing resource-intensive repair or replacement.

High-performance materials such as fibre-reinforced concrete, weathering steel, and engineered timber improve resistance to moisture, corrosion, and temperature fluctuations. Similarly, structural systems—like seismic framing, hurricane-resistant roofing, and flood-resistant foundations—help minimize damage during extreme events.

Moreover, durability enhances environmental performance by extending the lifespan of buildings. This spreads the embodied carbon—emissions from material production and construction—over more years, thereby reducing annualized impact. It also decreases the need for resource-intensive new construction (STRATA GLOBAL, 2024).

Adaptability

Beyond withstanding external forces, resilient architecture must also be adaptive, responding to changing environmental, social, and functional needs over time. Adaptability enables buildings to remain relevant over time, whether by accommodating new uses, responding to climatic variability, or integrating emerging technologies. At the spatial level, adaptable design is achieved through flexible layouts and modular construction techniques that allow buildings to be reconfigured with minimal intervention. This flexibility supports extended building lifespans and reduces the environmental costs of demolition and reconstruction. For example, open-plan office spaces can be converted into residential units, enabling cities to meet shifting housing demands while minimizing urban sprawl and maximizing the utility of existing infrastructure (Carbon Leadership Forum, 2020). Technological adaptability is increasingly enabled by smart building systems that integrate, for example, Artificial Intelligence (AI) and IoT sensors. These systems provide real-time data on occupancy, energy use, indoor air quality, and structural performance, allowing buildings to dynamically optimize their operations. Predictive maintenance, powered by machine learning algorithms, can identify inefficiencies and pre-empt failures, lowering operational costs and enhancing energy efficiency (Archisoup, 2024). Complementing these active systems, passive design strategies such as dynamic façades, self-shading elements, and ventilated envelopes allow buildings to adapt naturally to environmental fluctuations. These approaches collectively improve thermal comfort and reduce reliance on energy-intensive mechanical systems.

Sustainability

While closely related to resilience, sustainability emphasizes minimizing environmental harm over a building's lifecycle. It supports resilience indirectly by reducing dependency on resource-intensive systems.
Beyond material choices, energy efficiency is another fundamental aspect of sustainable design. Passive design strategies, including high-performance insulation, natural ventilation, and daylight optimization, can significantly reduce a building's reliance on mechanical heating and cooling systems. Research indicates that energy-efficient buildings can lower operational emissions by up to 40% compared to conventional designs (Myint & Shafique, 2024). Furthermore, integrating renewable energy sources, such as solar panels, and geothermal heating, can make buildings net-zero or even net-positive, producing more energy than they consume.

By integrating sustainable materials, energy-efficient design, and resource-conscious construction methods, buildings become more environmentally responsible.

IMPORTANCE OF DATA-DRIVEN DESIGN IN RESILIENT ARCHITECTURE

To address the objectives mentioned in the previous paragraphs, data driven approaches and algorithms can be used to enhance resilient architectural designs.

Overview

Data driven design describes the usage of real-world data to guide informed decision making, as opposed to using e.g. intuition or personal experience. In architecture, this can include data on:

- Energy use (e.g. heating, cooling)
- Environmental data (e.g. sunlight)
- User behavior (e.g. occupancy)

This data can be sorted and analyzed -often using digital and computational tools- and used as information in the design process to create a better-performing building while increasing user comfort (Ways Data-Driven Designs Improve Energy Efficiency, 2024). The rise of big data necessitates interdisciplinary methods for managing complex environmental, technical, and behavioral datasets. Data-driven design aims to use data inputs from the very beginning, increasing the need for tools and methods to deal with such complex issues as creating a sustainable built environment (Sunguroğlu Hensel et al., 2022).

A data-driven design process can support optimizing buildings to:

- Adapt to local conditions
- Increase energy efficiency
- Reduce waste

The next step is to identify different areas of application for the use of data-driven models and analyses designed to increase the sustainability and resilience of buildings. For a first assessment, the UN Sustainable Development Goals (SDG) are utilized to evaluate the sustainability impact of the techniques.

Environmental Risk Management

Environmental Risk Management, short ERM, describes an interdisciplinary process focused on identifying, evaluating, and reducing environmental risks to natural and built systems (Gomes et al., 2024). This Approach resonates with the UN sustainability goals, addressing topics like biodiversity loss, climate change and sustainable cities (United Nations, n.d.).

Data is integrated into ERM by collecting, processing, and analyzing information from multiple sources such as remote sensing (e.g. satellite imagery), Geographic Information Systems (GIS), sensor networks (e.g. weather stations) and historical records (e.g. flood maps, earthquake data). Additionally, emerging technologies such Building Information Modeling (BIM) and machine learning offer new possibilities for integrating diverse environmental and structural datasets.

In the field of building Science, ERM is applied to enhance the resilience and safety of a structure by systematically addressing environmental hazards throughout the planning, construction and operational phases. Site-specific threats such as floods, extreme wind events and heatwaves can be identified, and the information used for urban planning strategies, structural design and material selections (Rathnasiri et al., 2023).

BIM models can integrate GIS-based flood or seismic risk maps, which allows the targeted use of reinforcing measures or a more optimized design of the structure to be built. Furthermore, the incorporation of machine learning and AI applications into BIM can facilitate the automation of risk identification and prediction procedures (Boje et al., 2020).

Energy Performance Analysis

This type of data-driven analysis involves evaluating the energy consumption, efficiency and overall sustainability of a building through different stages of a building's life. It typically uses simulation models to predict how different material selections, design choices and operational strategies impact energy use. The utilization of energy performance analysis can help to optimize parameters like heating, cooling, ventilation and lighting systems and support planning strategies (Attia et al., 2012). When comparing this analysis method with the UN SDG's, it resonates with the goals affordable and clean energy, sustainable cities and responsible consumption, minimizing energy use and promoting sustainable design choices (United Nations, n.d.).

Advanced simulation tools like EnergyPlus, OpenStudio and Matlab are commonly used in the context of performance analysis and building science and can be integrated into BIM environments. BIM platforms, particularly those that employ Industry Foundation Classes (IFC), can facilitate semi-automated data transfer to a simulation software. This process reduces human error and optimises workflow, enhancing efficiency and accuracy in the process, as well as evidence-based decision-making (Bazjanac, 2008). As building science moves towards designing zero-energy buildings, aiming to generate as much energy locally from renewable sources as they consume over a year, highly efficient building envelopes and building systems are necessary to ensure resilience and user comfort. (Attia et al., 2012).

Lifecycle Assessment

Life cycle assessment (LCA) is a method to quantify environmental impacts connected to all phases of a product's life cycle. This ranges from the extraction of the raw material to production and use, up to disposal or recycling at the end of the cycle.

In architectural practice, LCA captures key metrics such as resource depletion, waste generation, acidification potential, and embodied carbon emissions (Cabeza et al., 2014). This whole-life approach aligns with sustainability goals such as responsible consumption, sustainable urban development, and climate action (United Nations, n.d.).

In building science, LCA is used in multiple applications, such as building certification, material selection, and design optimization. By analyzing a range of different design options in an early stage of the project, life cycle assessment methods can help choose systems and materials that minimize the negative environmental impact over the building's life cycle (Cabeza et al., 2014). Stakeholders can compare different impacts of e.g. structural systems - made from materials such as timber, steel or concrete - and decide in favour of low-carbon options. This approach can inform decision-making. LCA can be integrated into BIM workflows, providing feedback when realising different design options and helping to strike a balance between environmental performance and functionality (Wang et al., 2024). Additionally, lifecycle impact reporting is a requirement in achieving different sustainability certifications like BREEAM (UK) or LEED (USA) (Cabeza et al., 2014).

CONCLUSION

Construction projects as multifaceted problems are increasingly complex. The three pillars of sustainability, durability, and adaptability can be used as guidance for future projects to ensure a resilient built environment. By embedding these principles from the earliest stages of design, architects and planners can anticipate environmental risks, respond to changing user needs, and reduce the ecological impact of construction activities. The use of data-informed design methods -including the prediction of environmental risks, energy performance simulations and lifecycle assessments- strengthens the ability to make informed, future-orientated decisions. By integrating real-world data into the design and operational phases, buildings can become smarter, more efficient, and better prepared for future challenges.

Ultimately, combining traditional resilient strategies with emerging digital technologies creates opportunities for more durable, adaptive and sustainable architectural solutions. Future research and practice should continue to explore interdisciplinary approaches to ensure that the built environment not only responds to change but actively contributes to a more resilient and sustainable world.

REFERENCES

American Institute of Architects. BUILDINGS THAT LAST: DESIGN FOR ADAPTABILITY, DECONSTRUCTION, AND REUSE. (n.d.). Retrieved 8 April 2025, from https://content.aia.org/sites/default/files/2020-03/ADR-Guide-final_0.pdf.

Archisoup. (2024, June 24). Exploring Adaptive Architecture: Flexibility, Responsiveness, and Sustainability - archisoup. archisoup | Architecture Guides & Resources. Retrieved April 8, 2025, from https://www.archisoup.com/adaptive-architecture.

Connelly, E. B., Allen, C. R., Hatfield, K., Palma-Oliveira, J. M., Woods, D. D., & Linkov, I. (2017). Features of resilience. Environment Systems and Decisions, 37(1), 46–50. https://doi.org/10.1007/s10669-017-9634-9.

Architecture 2030. (n.d.). Embodied Carbon Actions – Architecture 2030. Retrieved April 8, 2025, from https://www.architecture2030.org/embodied-carbon-actions.

Attia, S., Gratia, E., De Herde, A., & Hensen, J. L. M. (2012). Simulation-based decision support tool for early stages of zero-energy building design. Energy and Buildings, 49, 2–15. https://doi.org/10.1016/j.enbuild.2012.01.028.

Bazjanac, V. (2008). IFC BIM-Based Methodology for Semi-Automated Building Energy Performance Simulation. Proceedings of the

25th International Conference on Information Technology in Construction.
Boje, C., Guerriero, A., Kubicki, S., & Rezgui, Y. (2020). Towards a semantic Construction Digital Twin: Directions for future research. Automation in Construction, 114, 103179. https://doi.org/10.1016/j.autcon.2020.103179
Bulman, J. (2024). What is Resilient Architecture? https://neumannmonson.com/blog/what-is-resilient-architecture.
Cabeza, L. F., Rincón, L., Vilariño, V., Pérez, G., & Castell, A. (2014). Life cycle assessment (LCA) and life cycle energy analysis (LCEA) of buildings and the building sector: A review. Renewable and Sustainable Energy Reviews, 29, 394–416. https://doi.org/10.1016/j.rser.2013.08.037.
Carbon Leadership Forum. (2020). Embodied carbon in buildings. https://carbonleadershipforum.org/embodied-carbon-101/.
Gomes, A., Islam, N. M., MS in Information Technology, Specialization- Data Management and Analytics, Washington University of Science and Technology, USA, Karim, M. R., & Business Analyst, Contract Manufacturing NE LLC, USA. (2024). Data-Driven Environmental Risk Management and Sustainability Analytics. Non Human Journal, 1(01), 100–113. https://doi.org/10.70008/jmldeds.v1i01.46
Myint, N., & Shafique, M. (2024). Embodied carbon emissions of buildings: Taking a step towards net zero buildings. *Case Studies in Construction Materials*, *20*, e03024. https://doi.org/10.1016/j.cscm.2024.e03024
Rathnasiri, P., Adeniyi, O., & Thurairajah, N. (2023). Data-driven approaches to built environment flood resilience: A scientometric and critical review. Advanced Engineering Informatics, 57, 102085. https://doi.org/10.1016/j.aei.2023.102085
STRATA GLOBAL. (2024, December 18). Understanding durability in construction: key factors and materials. Retrieved April 8, 2025, from https://www.strataglobal.com/glossary/durability/.
Sunguroğlu Hensel, D., Tyc, J., & Hensel, M. (2022). Data-driven design for Architecture and Environment Integration. SPOOL, 9(2), 19–34. https://doi.org/10.47982/spool.2022.1.02
United Nations. (n.d.). THE 17 GOALS | Sustainable Development. Retrieved 6 April 2025, from https://sdgs.un.org/goals
Wang, R., Abdul-Samad, Z., & Salleh, H. (2024). Systematic literature review of BIM application in the whole life cycle of green building projects. Advances in Building Energy Research, 18(5), 495–532. https://doi.org/10.1080/17512549.2024.2383866.
Ways Data-Driven Designs Improve Energy Efficiency. (2024, December 30). https://neuroject.com/data-driven-designs/

Generative Building Information Modeling

An overview of current usage and research

Günes Aydar[1], Laura Müllner[2], Kinga Średzińska[3]
[1,2,3]Technische Universität Wien
[1,2,3]{e0627278|e11701293|e12410042}@student.tuwien.ac.at

Generative Building Information Modeling combines the benefits of Generative Design, particularly suitable for early design phases, with BIM; resulting in data-informed yet quick generation of variations for iterative design stages. This allows for a continuous optimization of predetermined design goals based on selected fitness functions. Due to the wide range of applications across different disciplines and the variety of programs and workflows already in place, the integration of Generative BIM is still highly individualized both in the tools used as well as the intended results. Research on the topic focuses on improving fitness functions for specific use cases, supporting and streamlining the design process and improving the universality of developments for higher usability.

Keywords: *Building Information Modeling, Generative Design, Generative BIM, Algorithmic Planning, Design Automation*

INTRODUCTION

The need for time reduction and the enhancement of efficiency of design processes, along with the expectation of improved cost-effectiveness and sustainability in the built environment, underscore the necessity for automatization within the Architecture, Engineering and Construction industry. (Kookalani et al., 2024) The development of Building Information Modeling supports this shift by offering digital representations of a building's physical and functional characteristics while seamlessly integrating computational design, artificial intelligence, and optimization techniques into the typical design workflow within the AEC industry. (Ma et al., 2021) Complementing this, the development of generative design introduces algorithm-driven and computational methods that can automatically produce a wide range of design solutions based on predefined goals and constraints. While generative design primarily focuses on geometric modeling, its integration with BIM's rich informational attributes significantly enhances the constructability of the generated solutions. This enables automated and rapid design explorations, improving the buildability of generative outputs and extending BIM's capabilities into the early stages of design. (Kookalani et al., 2024; *What is Generative Design | Tools Software | Autodesk*, n.d.)

Generative BIM represents a paradigm shift and a significant evolution in automated design processes for the AEC industry. Combining the strengths of both technologies to facilitate rapid iterations, adaptive design responses, and data-driven optimization. Furthermore, GBIM addresses the challenges of the industry by providing higher levels of automation, which ultimately enhances decision-making, reduces design errors, and optimizes resource utilization, supporting more intelligent and sustainable design practices. (Ma et al., 2021)

This paper provides an overview of the current research and state of Generative BIM, highlighting key tools, challenges, and research directions. By synthesizing recent studies and case examples, the role of GBIM in shaping the future of digital construction is explored, addressing its potential benefits, limitations and outlook for future opportunities and research directions.

DEFINITION

Generative Building Information Modelling (GBIM) describes the combination of Generative Design and Building Information Modelling to take advantage of benefits while simultaneously offsetting weaknesses.

GBIM integrates the algorithmic design methods of generative design, intelligent design exploration and optimization capabilities with BIM's strengths in information modeling and the automated generation of construction data. This approach aims to facilitate the constructability of GD's automated solutions, improve BIM's capabilities in early design phases, and enable compliant, multi-objective design optimization and visualization. (Kookalani et al., 2024; Wang et al., 2010) By leveraging parametric modeling, computational design, and artificial intelligence to generate, it evaluates and refines multiple design alternatives based on predefined criteria and constraints.

EMERGENCE AND DEVELOPMENT

Early computational design and rule-based systems are conceptually rooted back in the 1970s. At first GD development focused on explicit programming, where human design knowledge was encoded into algorithms such as shape grammars, constraint programming, cellular automata, swarm intelligence and rule-based algorithms. (Jaisawal and Agrawal, 2021; Singh, no date)(Ma et al., 2021) True integration and practical application within AEC workflows accelerated with the rise of BIM platforms and the development of more accessible programming methods. (Coulthard (Wang, 2022; *From Design Automation to Generative Design in AEC | Autodesk University*, no date) Although the initial research on BIM, which back then was mostly concerning parametric modelling, leads back to the 1970s, the practical implementation of BIM tools within AEC industry is dated only to 2000s. (Ma et al., 2021)

Evolutionary and more specifically generative algorithms parametric design tools such as Grasshopper for Rhino laid the foundation for generative workflows by enabling designers to manipulate form and data through rule-based systems. (Ma et al., 2021)

Further developments involved the integration of GD techniques with BIM, which often was facilitated by Visual Programming Languages like Grasshopper and Dynamo, operating within or connected to BIM software such as Revit and ArchiCAD. The introduction of visual scripting platforms further empowered professionals to automate tasks and develop responsive, algorithm-driven models. (Jezyk and Kron, 2015)

Current trends move towards the incorporation of Artificial Intelligence, alongside advanced optimization techniques such as Topology Optimization in other to further enhance design automation by implicit learning from data. This signifies a shift, where BIM is no longer a documentation tool, but an intelligent design partner. (Ma et al., 2021)

CURRENT STATE

Both BIM and GD's usage as well as implemented workflows vary widely between different offices, disciplines and even regions. (Hamma-adama and Kouider, 2019) While workflows of, for example, architectural or structural planning processes had in many cases already included elements of both domains, the unification in the shape of GBIM is not least due to its emergent nature very diverse.

Disciplines of Application

The automatic rule-based generation of models allows for convenient option generation of facades, floorplan layouts or building envelope variations often employed during conceptual architectural planning phases. Based on the integrated information about materials, position and orientation, size or other architectural parameters within BIM models the evaluation and thus optimized selection of ideal design proposals becomes possible. Research investigates possible and current applications for automation of structural design aspects (Hamidavi et al., n.d.; Henríquez et al., n.d.), building physics applications including thermal, acoustic, life cycle and energy performance (Bhilare, n.d.; Ma & Wang, n.d.; Negendahl, n.d.; Agirbas, n.d.), optimization of MEP planning (Hu, n.d.; Pestana et al., 2024; Vahdatikhaki, 2022) as well as fire safety engineering (Lovreglio et al., 2022).

Depending on the discipline and specific task at hand, information and data points are extracted from the BIM-model, design task and expected user demand to construct fitness functions based on which iterative design options are evaluated.

Generative optimization of the site layout can utilize given inputs of site boundary, topological conditions, respective climate data as well as given building requirements to fulfil topological, economic or environmental design targets. Interior design optimization focuses on either idealized layout typology and room organization or adherence to performance parameters such as natural light, ventilation or user preferences. Loads, cost and layout considerations are balanced with structural behaviour and material requirements in static civil engineering applications. Mechanical, Electrical and Plumbing (MEP) optimization involves the gathering of input data, such as building geometry, expected loads, systems and uses and outputs automated installation placement, sizing or maximizes energy efficiency. Exterior or façade optimization tasks often involve a combination of both aesthetic and functional parameters. Window placement and sizing, materiality, orientation and geometric variations all influence building physics performance, daylight maximization and aesthetic quality simultaneously and may require the formulation of intermediate fitness parameters that can be quantified and optimized for. (Zhang and Zhang, 2025)

Generative BIM Tools

To join Generative Design and BIM into one workflow the specialized tools for each part are combined. As both areas have undergone rapid and varied development in recent years and established themselves in design and planning offices, their unification occurs in the form of many software-programming language combinations. (Ma *et al.*, no date)

Specifically the choice between textual or visual programming language can affect the interoperability and set the difficulty for entry for designers. Some of the most frequently covered implementations of Generative BIM covered by research are Autodesk Revit with Dynamo or McNeel's Rhinoceros with Grasshopper, as the provided visual programming interface reduces the need for actual programming experience for the development of the parametric model. (Chew, 2024)

Most of the commonly used Software employing a parametric design workflow, including Revit, Rhino/Grasshopper, ArchiCAD and Allplan, offer Application Programming Interfaces (APIs) with which textual programming languages can be used to customize the handling and modifying of data and model parameters. Particularly high-level programming languages such as Python or C# as well as custom languages like RhinoScript or SmartParts SL are employed to further extend the functionalities of the BIM software for the generative workflow. (Ma et al., 2021)

Workflows

According to Zhang and Zhang (2025), a typical generative BIM approach involves 4 stages: BIM information extraction, design requirement integration, generative method application and BIM conversion. In the first step, information specifically relevant to the optimization problem is extracted from the BIM model and subsequently supplemented with given design constraints and requirements in the second step. The gathered information acts as input and boundary conditions for the following generative step, in which a variety of algorithms may be employed to generate suitable design options. In the last step, the result is converted back into the BIM format for continued work.

Structural and MEP optimization applications often utilize an integrative approach within BIM applications such as Revit with the addition of task-specific algorithms. The established BIM-workflows within these disciplines allow for data extraction from the given architectural models based on which relevant parameters such as building component position, structural layouts as well as room sizes can be extracted. This allows for a largely automated determination of target parameters for the optimization process, including expected structural and energetic loads, environmental factors and possible clashes with architectural elements. The approach of utilizing integrated visual programming tools or APIs within existing BIM authoring software simplifies the initial data extraction and reimplementation within the model significantly. (Zhang and Zhang, 2025)

Optimization problems that lend themselves well to two- or lower dimensional abstractions or with less established integraative BIM workflows, such as interior layout and floorplan optimization, are often generated based on information extracted from BIM-models, but subsequently transformed into simpler data formats such as raster or vector images, simplified geometry or graph nodes. (Gan, 2022; Wang *et al.*, 2023) The generation and optimization process utilizes selected algorithms, such as genetic algorithms or machine learning approaches and may take place completely separately from the BIM environment. After a suitable design variant has been concluded, another process, depending on the resulting data, has to be employed to integrate it back into the BIM model, where further analysis and work can be done. (Quispe and Ulloa, 2021; Zhang and Zhang, 2025)

RESEARCH

Current research focuses on developing new systems and environments to better support design processes with the aim of improving the portability of GD-BIM development for higher universality. Limited case studies attempt to optimize parameters for fitness functions in specific cases. (He et al., 2023)

Areas of Research

A prominent area of research involves the integration and refinement of Deep Generative Models, such as Generative Adversarial Networks, Variational Autoencoders, and more recently, diffusion models, to learn complex design patterns. (He et al., 2023)

Research is actively exploring the synergy between these AI-driven generative techniques and Topology Optimization to create structurally efficient and materially optimized forms. (Mueller and Mikelsons, no date) Another area concerns the incorporation of generative AI tools, such as ChatGPT and Veras, into parametric modeling and BIM. The aim for this integration is to facilitate rapid exploration of design ideas and enhance collaboration between architects and AI systems, leading to more efficient and creative design processes (Ko et al., 2023)

Methods

In terms of research methodologies, current studies employ a variety of approaches to integrate generative design with BIM. One of them is algorithm development and application which involves the utilization and advancement of DGMs as well as the continued application of established algorithms like Genetic Algorithms, Shape Grammars, and Agent-Based Modeling. (Abrishami et al., 2020)

Moreover, researchers employing optimization techniques which methods ranging from gradient-based and gradient-free optimization to specialized approaches and multi-objective evolutionary algorithms, often integrated with machine learning-based surrogate models for efficiency. (Salah et al., 2020)

Additionally, researchers are exploring the use of prompt-based virtual assistant frameworks that integrate BIM with large language models to support natural language-based information. Research validation commonly involves computational experiments, applying the developed methods to architectural or engineering case studies. (Zheng and Fischer, 2023)

Software development is focused mostly on development of custom scripts or plugins using Visual Programming Languages like Dynamo or Grasshopper, or Textual Programming Languages like Python, to implement and test their generative workflows within existing BIM platforms. (Ma et al., 2021)

LIMITATIONS AND OUTLOOK

Usage of both Generative Design as well as emergent data-focused BIM workflows often requires programming experience on top of industry-specific knowledge, resulting in a very high barrier for entry. With the emergence of visual programming tools such as Revit Dynamo and Grasshopper for Rhino these requirements are no longer as strict, the complexity of constraint formulation, model configuration and weighting of fitness parameters however still requires extensive expertise and project-specific time investments. (Chew, 2024; Zhang and Zhang, 2025; Ma *et al.*, no date)

The precise evaluation of and computational effort associated with proposed GBIM workflows is directly linked with the amount of variables and thus generative model complexity. As such, much of the current research restricts the investigated parameters by only attempting to optimize one or two aspects at once. (Ma and Wang, no date) As design and planning processes have always been characterized by the specific need to account for many aspects simultaneously, future approaches are likely to focus on multi-objective optimization as well as their feasible implementation regarding computational and time investments.

To ensure that results of generative BIM processes meet user and functional requirements, "human-in-the-loop" or cooperative design approaches might increase in relevance, integrating human intuition and expertise within algorithmic design processes. (Chew, 2024; Zhang and Zhang, 2025)

Current issues stemming from the inherent complexity of input data, generative modeling approaches as well as setting and evaluation of fitness parameters may undergo rapid changes with the ongoing integration and evolution of machine learning and generative AI. (Ma and Wang, no date)

REFERENCES

Abrishami, S., Goulding, J., Rahimian, F., 2020. Generative BIM workspace for AEC conceptual design automation: prototype development. ECAM 28, 482–509. https://doi.org/10.1108/ECAM-04-2020-0256

Agirbas, A. (n.d.). _Multi-objective building design optimisation using acoustics and daylighting_.

Bhilare, S. et al. (2024a) 'BIM-Based Generative Design Processes for Targeted Value Design of Low Carbon Buildings', in S. Desjardins and G.J. Poitras (eds) Proceedings of the Canadian Society for Civil Engineering Annual Conference 2023, Volume 1. Cham: Springer Nature Switzerland (Lecture Notes in Civil Engineering), pp. 253–265. Available at: https://doi.org/10.1007/978-3-031-60415-7_19.

Cai, W. et al. (2024) '3D reconstruction of building interiors based on scan-to-BIM and generative design for as-built building', Engineering, Construction and Architectural Management [Preprint]. Available at: https://doi.org/10.1108/ECAM-07-2024-0984.

Chew, Z.X. (2024) 'Generative Design in the Built Environment', *Automation in Construction* [Preprint], (166). Available at: https://doi.org/10.1016/j.autcon.2024.105638 .

Coulthard, J. and Wang, C.J. (2022) 'Generative Design and Topology Optimisation of Products for Additive Manufacturing', in M. Shafik and K. Case (eds) Advances in ransdisciplinary Engineering. IOS Press. Available at: https://doi.org/10.3233/ATDE220570.

From Design Automation to Generative Design in AEC | Autodesk University (no date). Available at: https://www.autodesk.com/autodesk-university/article/Design-Automation-Generative-Design-AEC-2020 (Accessed: 2 April 2025).

Gan, V.J.L. (2022) 'BIM-based graph data model for automatic generative design of modular buildings', *Automation in Construction*, 134, p. 104062. Available at: https://doi.org/10.1016/j.autcon.2021.104062 .

Hamidavi, T., Abrishami, S., & Hosseini, M. R. (n.d.). _Towards intelligent structural design of buildings: A BIM-based solution_.

Hamma-adama, M. and Kouider, T. (2019) 'Comparative Analysis of BIM Adoption Efforts by Developed Countries as Precedent for New Adopter Countries', *Current Journal of Applied Science and Technology*, pp. 1–15. Available at: https://doi.org/10.9734/cjast/2019/v36i230224.

He, Z., Wang, Y.-H. and Zhang, J. (2025) 'Generative AIBIM: An automatic and intelligent structural design pipeline integrating BIM and generative AI', Information Fusion, 114, p. 102654. Available at: https://doi.org/10.1016/j.inffus.2024.102654.

Henríquez, D., Herrera, R. F., & Vielma, J. C. (n.d.). _Method for Designing Prequalified Connections Using Generative Design_.

Hu, Z.-Z. (n.d.). _BIM-based integrated delivery technologies for intelligent MEP management in the operation and maintenance phase_.

Jaisawal, R. and Agrawal, V. (2021) 'Generative Design Method (GDM) – A State of Art', *IOP Conference Series: Materials Science and Engineering*, 1104(1), p. 012036. Available at: https://doi.org/10.1088/1757-899X/1104/1/012036.

Jezyk, M. and Kron, Z. (2015) *Q&A about Dynamo*. Available at: https://dynamobim.org/qa-about-dynamo (Accessed: 3 April 2025).

Kensek, K.M. and Noble, D.E. (2014) Building information modeling: BIM in current and future practice. Hoboken, NJ: Wiley.

Kookalani, S. et al. (2024) 'Trajectory of building and structural design automation from generative design towards the integration of deep generative models and optimization: A review', Journal of Building Engineering, 97, p. 110972. Available at: https://doi.org/10.1016/j.jobe.2024.110972.

Lovreglio, R., Thompson, P., & Feng, Z. (2022). Automation in Fire Safety Engineering Using BIM and Generative Design. _Fire Technology_, _58_(1), 1–5. https://doi.org/10.1007/s10694-021-01153-7

Liu, H. et al. (2021) 'Design for Manufacturing and Assembly: A BIM-Enabled Generative Framework for Building Panelization Design', Advances in Civil Engineering. Edited by J. Zhou, 2021(1), p. 5554551. Available at: https://doi.org/10.1155/2021/5554551.

Ma, W. et al. (2021) 'Generative Design in Building Information Modelling (BIM): Approaches and Requirements', Sensors, 21(16), p. 5439. Available at: https://doi.org/10.3390/s21165439.

Ma, W., & Wang, X. (2025). BIM-based generative design approach for integral residential energy-efficient façades. _Energy and Buildings_, _328_, 115118. https://doi.org/10.1016/j.enbuild.2024.115118

Mueller, P. and Mikelsons, L. (no date) 'Exploring the Potentials and Challenges of Deep Generative Models in Product Design Conception'.

Negendahl, K. (n.d.). _Building performance simulation in the early design stage: An introduction to integrated dynamic models_.

Pestana, E., Paice, A., & West, S. (2024). Optimizing MEP design in early AEC projects through generative design. _Automation in Construction_, _165_, 105566. https://doi.org/10.1016/j.autcon.2024.105566

Quispe, L. and Ulloa, W. (2021) 'Application of BIM Tools in Parametric and Generative Design for the Conception of Complex Structures', IOP Conference Series: Materials Science and Engineering, 1203(2), p. 022070. Available at: https://doi.org/10.1088/1757-899X/1203/2/022070.

Salah, M., Elbeltagi, E. and Elsheikh, A. (2024) 'Optimization of construction site layout using BIM generative design', International Journal of Construction Management, 24(3), pp. 314–322. Available at: https://doi.org/10.1080/15623599.2023.2222997.

Singh, V. (no date) 'Towards an integrated generative design framework'.

Vahdatikhaki, F., Salimzadeh, N. and Hammad, A. (2022) 'Optimization of PV modules layout on high rise building skins using a BIM-based generative design approach', Energy and Buildings, 258, p. 111787. Available at: https://doi.org/10.1016/j.enbuild.2021.111787

Wang, J., Li, J. and Chen, X. (2010) 'Parametric Design Based on Building Information Modeling for Sustainable Buildings', in *2010 International Conference on Challenges in Environmental Science and Computer Engineering*. IEEE, pp. 236–239. Available at: https://doi.org/10.1109/CESCE.2010.285.

Wang, L. *et al.* (2023) 'Automated building layout generation using deep learning and graph algorithms', *Automation in Construction*, 154, p. 105036. Available at: https://doi.org/10.1016/j.autcon.2023.105036.

Zhang, H. and Zhang, R. (2025) 'BIM-Based Generative Design: A Comprehensive Review Across Key Domains', in A. Francis, E.

Miresco, and S. Melhado (eds) *Advances in Information Technology in Civil and Building Engineering*. Cham: Springer Nature Switzerland (Lecture Notes in Civil Engineering), pp. 153–166. Available at: https://doi.org/10.1007/978-3-031-84208-5_13.

Zheng, J. and Fischer, M. (2023) 'BIM-GPT: a Prompt-Based Virtual Assistant Framework for BIM Information Retrieval'. arXiv. Available at: https://doi.org/10.48550/arXiv.2304.09333

Is Machine Learning/Artificial Intelligence "algorithmic"?

Emir Softic[1]
[1] *Technical University of Vienna*
[1]*e.softic2001@gmail.com*

The question of whether machine learning and artificial intelligence models are truly algorithmic remains a subject of debate. The current advancements in information technology have reached a point where some problems can be solved without knowing why a certain algorithm made a certain decision, thereby bringing ethical, data management and security concerns to light. This paper explores differences between white-box and black-box models, emphasizing their transparency and interpretability, as well as mentioning explainable artificial intelligence as a potential solution to opaque results given by machine learning and artificial intelligence models.

Keywords: *black-box model, white-box model, Machine learning, Artificial intelligence*

INTRODUCTION

Early computers were built for complex calculations, storing both data and instructions. Over time, they began processing data based on structured models of the real world but still followed explicit commands. The next step was enabling computers to learn from data, extracting rules for classification and prediction, which raised machine learning and AI. A breakthrough, according to Cohen (2025), came with deep learning, inspired by brain-like networks of interconnected layers. AI is now transforming many fields, including biomedical research and clinical practice, at an accelerating pace. Unlike traditional information systems, AI algorithms can function as "black boxes." Previously, developers could fully explain how an algorithm worked and why it produced a given output, as it followed human-defined rules. AI applications, however, generate internal structures that determine outputs, but even programmers cannot always explain their decisions. Asatiani et al(2020) state that, information system (IS) researchers are working to enhance algorithm transparency, aiming to make AI decisions more understandable—a field known as "AI explainability." This paper aims to explore whether machine learning (ML) and artificial intelligence (AI) can be considered truly "algorithmic" by examining the distinctions between black-box and white-box models. It further investigates the role of explainable AI (XAI) in bridging the gap between these models, addressing the challenges of transparency and interpretability in AI decision-making.

Defining algorithms in the context of AI/ML

Traditional algorithms are fundamental to computer science and mathematics, providing systematic methods for solving problems and performing computations. Hill (2016), write that at their core, algorithms are well-defined sequences of steps designed to transform inputs into outputs through deterministic processes. The concept of an algorithm can also be extended beyond, as it can describe procedures in everyday tasks, such as following a recipe or solving a puzzle. A well-formed algorithm possesses several key characteristics well described in Geek for geeks post. It must be unambiguous, ensuring that each step is clearly defined and interpretable. It should accept zero or more well-defined inputs and produce at least one output. The algorithm must be finite, terminating after a limited number of steps. Moreover, it should be feasible, meaning it can be executed with available computational resources and language-independent, allowing implementation in various programming languages without altering its fundamental logic.

Different types of algorithms address a wide range of computational problems which were summed up in Wikipedia. Brute-force algorithms provide straightforward but inefficient solutions, while recursive algorithms solve problems by breaking them into smaller subproblems. Backtracking algorithms systematically explore all possible solutions, discarding incorrect paths. Searching and sorting algorithms are fundamental to data manipulation, whereas hashing techniques enable efficient data retrieval and so on. The necessity of algorithms arises from their role in automating and optimizing complex computations. They enable computers to process vast amounts of data efficiently, reducing human effort and minimizing errors. In practical applications, they enhance decision-making, streamline industrial processes, and contribute to advancements in fields such as healthcare, finance, and engineering. Despite their advantages, algorithms have limitations. Designing efficient algorithms requires significant time and expertise. Complex algorithms can be difficult to understand and implement, and some computational problems remain intractable even with the most advanced methods. Furthermore, certain algorithmic tasks, such as those involving artificial intelligence, rely on heuristic approaches (approximate solutions when exact methods are too complex or computationally expensive) rather than purely deterministic logic. Pseudocode af an algorithm is a simplified, informal way of representing an algorithm in human-readable language, often without the strict syntax of programming languages. They are used to help us focus on the logic and steps of an algorithm without worrying about specific programming details, as shown in the example below.

```
Algorithm: FindMax
Set max = list[0]
For each number in list starting from index 1:
   If number > max:
      Set max = number
Return max
```

Machine Learning (ML) has emerged as a transformative technology, fueled by advancements in data availability, algorithms, and computational power. ML, particularly through Deep Learning (DL) and neural networks, allows the resolution of complex problems where traditional algorithms fall short. Its applications span from voice assistants (e.g., Siri, Alexa) to critical sectors like healthcare, autonomous vehicles, and credit scoring. However, despite its benefits, the slow adoption of ML in business is primarily due to concerns about model quality and reliability. These models, especially neural networks, can struggle when exposed to new or slightly different data, raising safety issues that are unacceptable in applications like healthcare and autonomous driving. A key distinction in the AI field is the difference between algorithms and ML models. While algorithms are, as described, predefined sets of steps designed by humans to solve problems, ML models are built by training algorithms on data to make predictions or decisions. Unlike traditional programs that follow fixed instructions, ML models learn from data, making them flexible but, on the other hand, opaque. This contrast is crucial in understanding how AI systems function: algorithms are based on deterministic, predefined rules, whereas ML models evolve based on data input, and their internal workings may not be easily interpretable. AI systems can be categorized into three primary types: Knowledge-based or heuristic algorithms, which use explicit representations; Evolutionary or metaheuristic algorithms, which optimize solutions through trial-and-error methods; and Machine Learning algorithms, which learn from data. These categories enable a wide range of applications, from supervised learning, where models predict outcomes based on labeled data, to unsupervised learning, which uncovers hidden patterns in data. The transparency and explainability of AI models are essential for fostering trust. While algorithms offer clear and predictable outputs, the flexibility of ML models makes them harder to interpret. The distinction between static algorithms and adaptive ML models underscores the shift from human-driven problem-solving to data-driven learning, which is pivotal for domains like autonomous vehicles, healthcare, and financial forecasting. To ensure the safe and ethical deployment of these systems, it is crucial to focus on improving fairness, transparency, and the interpretability of AI models.

Black-box vs. White-box models

- White-box models

White-box machine learning models are distinguished by their transparency, simplicity, and ease of interpretation, making them highly valuable in fields where understanding the decision-making process is essential. These models allow users to see exactly how input data is processed, which variables are prioritized, and how conclusions are drawn according to the active state publishment. Common examples include linear models, decision trees, and regression trees, all of which provide clear, logical structures that are easy to follow. Decision trees, for instance, function as directed graphs, where each node represents a decision criterion based on input features, leading to an outcome or classification. Their ability to visually map the decision-making process makes them particularly useful for experts who need to validate predictions. Unlike black-box models, which require additional techniques to explain their behavior, white-box models offer built-in interpretability, allowing direct insights into their internal mechanics without needing external aids. This clarity makes them ideal for scenarios where human oversight and accountability are necessary. In high-risk applications, such as medical diagnosis or loan approvals, being able to trace the reasoning behind predictions is crucial for ensuring fairness and reducing bias. Additionally, businesses benefit from white-box models by gaining insights into what drives predictions, allowing them to refine processes and improve decision-making strategies. These models are also favored for regulatory compliance, as they enable organizations to meet legal and ethical requirements by providing clear explanations of their decision-making processes. Furthermore, white-box models can be robust and flexible, handling both categorical and numerical data, and they can often deal effectively with missing values. While simple models like single decision trees may have limited predictive power, ensemble techniques such as random forests can improve performance while maintaining a degree of interpretability. However, despite their advantages, white-box models come with notable limitations discussed by Loyola-Gonzalez (2019), particularly when handling complex, non-linear relationships. Their reliance on straightforward logic makes them less effective for capturing intricate patterns in high-dimensional datasets, where deep learning or more sophisticated techniques are needed. Decision trees, for example, on become unstable, as small variations in training data may lead to significantly different structures, reducing reliability. Additionally, while individual decision trees are easy to interpret, ensembles of trees, such as those in random forests, can introduce complexity, making it harder for experts to fully grasp the decision process. Another drawback is that white-box models often do not achieve the same level of predictive performance as black-box models, particularly in scenarios requiring advanced pattern recognition, such as image processing or natural language understanding. As a result, while white-box models excel in transparency, accountability, and ease of validation, they may not be suitable for tasks demanding the highest levels of predictive accuracy and complexity.

- Black-box models:

Black-box machine learning models based on Koshiyama et al (2024), such as neural networks, support vector machines (SVMs), k-nearest neighbors (kNN), deep learning models, boosting models, and random forests, are all powerful tools for solving complex machine learning tasks [8]. These models are known for their ability to capture intricate, nonlinear relationships in large datasets, but this comes at the cost of transparency and interpretability. Unlike traditional models, which offer more straightforward decision-making processes, black-box models operate through multiple layers of nonlinear algorithms and transformations that are inherently difficult for humans to understand. Even if the internal workings were accessible, their complexity would still make them difficult to interpret. For instance, deep learning models use neural networks with hidden layers that process data in ways that are not easily traceable, while random forests and boosting models rely on ensemble learning, combining multiple decision trees that make their decision-making non-transparent. Additionally, techniques like embeddings and hidden representations create features beyond human comprehension, adding to the models' complexity. The primary advantage of black-box models, based on the writing of Damyanova (2020), is their impressive predictive accuracy. These models excel in handling large, complex datasets with nonlinear relationships, which allows them to make accurate predictions in tasks like image recognition, speech processing, natural language processing, recommendation systems, and fraud detection. Their ability to learn from vast amounts of data enables them to capture patterns that simpler models cannot, making them invaluable for applications such as generative models (e.g., GPT, used in ChatGPT and text-to-image generators like Stable Diffusion). Their success lies in modeling real-world behaviors with deep, nonlinear interactions between data points, closely mimicking how humans process information. Despite their high performance, black-box models come with significant challenges, which were discussed in the post of Advsolned. The most notable issue is the lack of interpretability, making it difficult to explain how a model arrived at a specific decision. This lack of transparency is particularly problematic in high-stakes fields like healthcare, autonomous driving, and finance, where understanding a model's reasoning is crucial for trust and accountability. It also complicates debugging and refining the model, as pinpointing errors or making necessary adjustments is not straightforward. Furthermore, the results of black-box models may not always be reproducible. Due to their complexity and the many factors influencing their predictions, these models can yield inconsistent results when applied to different datasets or even when the same dataset is processed multiple times, posing a challenge for their use in critical applications where predictable outcomes are essential. Thus, while black-box models continue to drive innovation, their application must be carefully managed to balance performance with the need for transparency and trust. This can be visualized via Figure 1.

Explainable Artificial Intelligence

Explainable AI (XAI) is a field within artificial intelligence that aims to make AI decision-making processes transparent and understandable to humans. The main objectives of XAI are to improve trust, ensure fairness, and enhance model accountability by providing clear, human-readable explanations of how and why AI models arrive at specific conclusions, which was written by insights organization. This is particularly crucial in high-stakes fields, where understanding AI decisions is vital for ethical considerations, regulatory compliance, and user trust. XAI helps prevent AI systems from being perceived as "black boxes," promoting a clearer understanding of their operations and fostering trust in AI-driven decision-making. XAI encompasses several techniques that aim to explain AI model decisions. These methods, according to Gerlings, Shollo, and Constantiou (2020), can be broadly categorized into post-hoc explanation methods and interpretable models. Post-Hoc explanation methods provide insights into the decision-making process of AI models after the decision has been made, without altering the underlying model. Two prominent post-hoc explanation techniques include SHapley Additive exPlanations and Local Interpretable Model-agnostic Explanations. SHAP (SHapley Additive exPlanations): This technique is grounded in cooperative game theory and assigns each feature a contribution value toward the model's final prediction. SHAP provides both global and local interpretability by explaining individual predictions based on the importance of each feature. It helps to determine the contribution of different features toward a specific outcome, enhancing the model's transparency, according to Van den Broeck et al (2022). LIME (Local Interpretable Model-agnostic Explanations): LIME approximates complex model behavior by training an interpretable model (such as a linear regression) on a subset of perturbed data near a specific instance. This method provides localized explanations, helping to explain why a model made a particular decision for a given input. LIME is model-agnostic based on writing of Zafar and Khan (2021), meaning it can be applied to any machine learning model, regardless of its underlying complexity. A paper by Gunning et al (2019) states that some models are inherently interpretable, meaning their structure allows for direct insight into decision-making. These models are designed to be transparent and easy to understand without the need for additional explanation techniques. Decision Trees are a simple and transparent model where decisions are made by traversing a flowchart-like structure. Each decision node represents a decision based on a feature, and the branches indicate the possible outcomes. This structure makes it easy to trace the reasoning behind the model's predictions, making decision trees highly interpretable. Linear Regression is another inherently interpretable model. The relationships between inputs and outputs are explicitly defined by linear equations, making it easy to understand how changes in the input variables affect the predictions. The coefficients of the model directly indicate the contribution of each feature to the final outcome. While XAI offers significant benefits, it also presents challenges. A key issue is the trade-off between accuracy and interpretability. Complex models, such as deep neural networks, tend to deliver high accuracy but are often difficult to interpret. On the other hand, simpler models like decision trees are more interpretable but may not achieve the same level of predictive performance. Striking a balance between these two aspects is essential depending on the specific application and the importance of model transparency versus accuracy. In some cases, hybrid approaches are employed, combining interpretable models with more complex ones, to enhance both performance and transparency. However, understanding and managing this trade-off remains a crucial aspect of the ongoing development of XAI technologies.

A Framework for Explaining Black-Box AI Systems

This framework provides a way to explain the behavior of black-box AI systems. Asatiani et al (2020) focus on six key elements. AI models range from simple, like decision trees, to complex, like deep neural networks, which are harder to interpret. Their goals are defined by measurable metrics, helping assess performance and uncover biases. Training data quality affects outcomes—biased data leads to biased decisions. Input data must be reliable, as poor data can cause errors, like in Australia's welfare debt case. Since AI outputs aren't always transparent, human oversight is needed to ensure accuracy and fairness.

References

Activestate - https://www.activestate.com/blog/white-box-vs-black-box-algorithms-in-machine-learning/

Advsolned - https://www.advsolned.com/an-ai-ml-model-is-not-an-algorithm/

Ali, S., Abuhmed, T., El-Sappagh, S., Muhammad, K., Alonso-Moral, J.M., Confalonieri, R., Guidotti, R., Del Ser, J., Díaz-Rodríguez, N. and Herrera, F., 2023. Explainable Artificial Intelligence (XAI): What we know and what is left to attain Trustworthy Artificial Intelligence. Information Fusion, 99, p.101805.

Asatiani, A., Malo, P., Nagbøl, P.R., Penttinen, E., Rinta-Kahila, T. and Salovaara, A., 2020. Challenges of explaining the behavior of black-box AI systems. MIS Quarterly Executive, 19(4), pp.259-278.

Cohen, S., 2025. The evolution of machine learning: Past, present, and future. In Artificial Intelligence in Pathology (pp. 3-14). Elsevier.

Damyanova, B., 2020. Quality attributes in AI-ML-based systems: differences and challenges (Bachelor's thesis).

Gerlings, J., Shollo, A. and Constantiou, I., 2020. Reviewing the need for explainable artificial intelligence (xAI). arXiv preprint arXiv:2012.01007.

Geek for geeks - https://www.geeksforgeeks.org/introduction-to-algorithms/

Gunning, D., Stefik, M., Choi, J., Miller, T., Stumpf, S. and Yang, G.Z., 2019. XAI—Explainable artificial intelligence. Science Robotics, 4(37), p.eaay7120.

Hill, R.K., 2016. What an algorithm is. Philosophy & Technology, 29, pp.35-59.

Insights - https://insights.sei.cmu.edu/blog/what-is-explainable-ai/

Koshiyama, A., Kazim, E., Treleaven, P., Rai, P., Szpruch, L., Pavey, G., Ahamat, G., Leutner, F., Goebel, R., Knight, A. and Adams, J., 2024. Towards algorithm auditing: managing legal, ethical and technological risks of AI, ML and associated algorithms. Royal Society Open Science, 11(5), p.230859.

Loyola-Gonzalez, O., 2019. Black-box vs. white-box: Understanding their advantages and weaknesses from a practical point of view. IEEE Access, 7, pp.154096-154113.

Van den Broeck, G., Lykov, A., Schleich, M. and Suciu, D., 2022. On the tractability of SHAP explanations. Journal of Artificial Intelligence Research, 74, pp.851-886.

Wikipedia - https://en.wikipedia.org/wiki/Algorithm

Zafar, M.R. and Khan, N., 2021. Deterministic local interpretable model-agnostic explanations for stable explainability. Machine Learning and Knowledge Extraction, 3(3), pp.525-541

IN-DEPTH PAPERS

The Act of Programming

Philip Grawatsch-Pollhammer
TU Wien, Vienna, Austria
philip.grawatsch-pollhammer@student.tuwien.ac.at

The act of programming, translating algorithmic ideas into executable instructions, represents far more than technical execution. This paper argues for a holistic understanding of programming as a discipline integrating systematic design, adaptable management, and rigorous quality assurance. It explores the essential journey from conceptualizing an algorithm to refining functional code, considering the roles of both textual and graphical language paradigms. Recognizing that creating complex software demands structure and flexibility, the discussion highlights the value of modern methodologies like Agile development in navigating uncertainty and fostering collaboration, contrasting them with traditional approaches. Essential to delivering reliable software is the critical interplay between development and testing, employing various levels of scrutiny to ensure correctness. This entire process culminates in addressing the fundamental goals of Verification – ensuring the software is built correctly according to specifications – and Validation – ensuring the right software is built to effectively meet user needs, key concepts in software quality assurance. The paper contends that successfully producing valuable software relies on the deliberate integration of these facets: process, management, testing, and goal-oriented evaluation, an understanding particularly pertinent as computational approaches increasingly shape fields like architectural planning.

Keywords: *Programming, Software Development Process, Agile Development, Software Testing, Verification and Validation.*

INTRODUCTION

The act of programming, at its core, is the process of translating an algorithm – a defined sequence of steps to solve a problem – into a precise language that a computer can interpret and execute. It forms the essential link between a conceptual solution and a functional computational tool. However, reducing programming solely to the mechanical task of writing code fails to capture the full scope of this activity. It is fundamentally an act of problem-solving, requiring logical thought, careful design, and structured implementation.

In contemporary society, programming extends its reach into nearly every domain, significantly impacting fields beyond traditional information technology. Within architectural planning, for example, the use of algorithms and custom programs is becoming increasingly prevalent. These tools allow designers to generate and explore complex forms, automate laborious drafting processes, simulate building performance, and manage intricate project information, thereby enabling new approaches to design and construction (Terzidis, 2006). Consequently, a clear understanding of the programming process is valuable for anyone engaging with these evolving digital workflows in architecture.

This paper explores the multifaceted nature of the "Act of Programming". The focus is on the practical journey involved in creating software, following the path from an initial idea through to its realization as a functional program. Furthermore, it examines how this creative process is structured and managed, particularly through modern methodologies like Agile development, which emphasize flexibility and collaboration. A crucial aspect of developing reliable software is ensuring its quality, leading to a discussion of the indispensable role of testing at various stages.

Ultimately, effective programming strives to satisfy two critical objectives, related to fundamental questions famously highlighted within software engineering by pioneers such as Boehm (1984): "Are we building the system right?" and "Are we building the right system?". These questions touch upon the core concepts of Verification and Validation, which are vital for ensuring a software project's success. These concepts, and how they relate to the programming act, will be explored in detail later in this paper. The aim is to investigate these key components – the development workflow, project management strategies, quality assurance through testing, and the guiding principles of verification and validation – to present a holistic view of the act of programming.

FROM IDEA TO INSTRUCTION: THE DEVELOPMENT JOURNEY

Creating a computer program is a systematic process that transforms an abstract idea or requirement into concrete, executable instructions. This journey involves several distinct stages, moving from high-level conceptualization to detailed implementation and refinement. Understanding these stages provides insight into the practical realities of the act of programming.

Conceptualizing the Solution: Designing Algorithms

Before any code is written, the first crucial step is to thoroughly understand the problem that the program is intended to solve. This involves clarifying requirements and defining the desired outcome. Once the problem is understood, the next step is to design a solution plan, known as an algorithm. An algorithm provides a logical, step-by-step sequence of actions required to achieve the desired result (Knuth, 1997). This design phase is critical; a flawed algorithm will lead to a flawed program, regardless of how well the code is written. Techniques such as writing pseudocode (a structured, informal description of the algorithm using natural language mixed with programming conventions) or creating flowcharts (graphical representations of the process flow) are often employed to develop and communicate the algorithmic logic before proceeding to implementation (Sommerville, 2011).

Language Matters: Textual and Graphical Programming

With a clear algorithm designed, the next step involves choosing an appropriate programming language to translate the algorithm into instructions the computer can execute. Programming languages act as the medium for this translation. They generally fall into two broad categories: textual and graphical.

Textual languages, such as Python, Java, or C++, require programmers to write instructions using specific keywords, symbols, and syntax rules, arranged sequentially line by line. These languages offer great flexibility and are widely used across a vast range of applications (Pratt and Zelkowitz, 2001).

Graphical languages, often called visual programming languages (VPLs), take a different approach. Instead of typing text, programmers typically manipulate visual elements, such as blocks or nodes representing operations or data, and connect them with wires or links to show the flow of execution or data (Burnett, 1995). Examples prominent in architectural design include Grasshopper (for Rhino 3D) and Dynamo (for Revit), which allow designers to build complex parametric models and automate tasks visually, often without extensive traditional coding experience (Tedeschi, 2014). The choice between textual and graphical languages often depends on the specific problem domain, the target platform, the required performance, and the programmer's familiarity and expertise.

The Core Task: Writing and Refining Code

This stage involves the actual translation of the designed algorithm into the chosen programming language's specific syntax. Often referred to simply as 'coding', this process demands careful attention to detail. However, writing effective code goes beyond merely achieving functionality. Good programming practice emphasizes writing code that is also readable, understandable, and maintainable – not only for potential collaborators but also for the original programmer revisiting the code later. This involves adhering to established coding conventions or style guides, using meaningful names for variables and functions, structuring the code logically, and incorporating comments to clarify the purpose or complexity of certain parts (McConnell, 2004). Writing code is frequently an iterative process; programmers often write segments of code, perform basic checks or tests to ensure they function as expected, and then refine or build upon them.

Identifying Flaws: The Role of Debugging

It is rare for even experienced programmers to write a program that works perfectly on the first attempt. Errors, commonly known as 'bugs', are an almost inevitable part of software development. Debugging is the systematic process of finding the root cause of these errors and correcting them (Zeller, 2009). Bugs can manifest in various ways: syntax errors occur when the code violates the language's grammatical rules and are typically caught before the program runs; runtime errors happen during program execution (e.g., attempting an impossible operation like dividing by zero); and logical errors are often the most challenging, where the program runs without crashing but produces incorrect or unexpected results due to flaws in the programmer's logic. Debugging involves a range of techniques, including using specialized software tools called debuggers that allow step-by-step execution, analysing program output, inserting temporary 'print' statements to track data values, and applying logical deduction to pinpoint the source of the error. Far from being a sign of failure, debugging is an essential skill and an integral, expected part of the software development lifecycle.

NAVIGATING COMPLEXITY: SOFTWARE DEVELOPMENT METHODOLOGIES

As software projects grow in scope and involve collaborative effort, simply following the basic programming steps is often not enough to ensure success. The inherent complexity, the need for team coordination, and the frequent possibility of changing requirements necessitate structured approaches for managing the development process. These approaches are known as software development methodologies or process models (Sommerville, 2011). Over time, different models have emerged, ranging from highly structured, sequential methods to more flexible, adaptive frameworks like Agile.

Traditional Approaches: The Waterfall Model

One of the earliest and most influential methodologies is the Waterfall model. Often associated with manufacturing processes, it organizes software development into a linear sequence of distinct phases: typically Requirements Definition, System Design, Implementation (Coding), Testing, Deployment, and finally, Maintenance. The defining characteristic is that each phase should be fully completed before the next one begins, creating a downward flow resembling a waterfall (Royce, 1987).

The Waterfall model places strong emphasis on comprehensive upfront planning and documentation for each phase. It can be suitable for projects where the requirements are exceptionally stable, clearly defined from the outset, and unlikely to change significantly – perhaps for developing a straightforward update to an existing, well-understood system. However, its primary drawback is its rigidity. Because phases are sequential and self-contained, accommodating changes or addressing misunderstandings discovered late in the process (for instance, during testing) can be very difficult and expensive, often requiring significant backtracking and rework (Pressman, 2020).

Embracing Change: The Agile Philosophy

Recognizing the limitations of rigid models, especially when dealing with the uncertainty and evolving requirements common in many software projects, a different philosophy emerged: Agile. This philosophy was consolidated in the 'Manifesto for Agile Software Development' (Beck et al., 2001). Instead of a single, rigid process, Agile emphasizes a set of guiding values and principles. The four core values are:

- Individuals and interactions over processes and tools
- Working software over comprehensive documentation
- Customer collaboration over contract negotiation
- Responding to change over following a plan

Key principles underlying Agile include iterative development (building software in small, repeated cycles), incremental delivery (releasing functional portions of the software frequently), close and continuous collaboration between developers and stakeholders (such as users or clients), and maintaining the flexibility to adapt to new requirements or insights throughout the project (Fowler and Highsmith, 2001). Agile fundamentally accepts that not everything can be known at the start and prioritizes adaptability.

Agile Practices in Action

Agile itself is an umbrella term covering various specific frameworks and practices that implement its philosophy. Scrum is arguably the most widely used Agile framework today. Scrum structures development work into fixed-length iterations, usually lasting between one and four weeks, known as Sprints. The goal of each Sprint is to produce a potentially usable increment of the software product (Schwaber and Sutherland, 2011).

Scrum utilizes specific roles, events (meetings), and artifacts (documents/tools) to manage the process:

- **Roles:** The Product Owner is responsible for defining features and prioritizing work from the customer's perspective. The Development Team builds the software. The Scrum Master acts as a facilitator, ensuring the team follows Scrum practices and removing obstacles.
- **Events:** These include Sprint Planning (selecting work for the upcoming Sprint), the Daily Scrum (a short daily synchronization meeting), the Sprint Review (demonstrating the results of the Sprint), and the Sprint Retrospective (reflecting on the process to identify improvements).
- **Artifacts:** The Product Backlog is the master list of all desired features. The Sprint Backlog contains the tasks chosen for the current Sprint. The Increment represents the sum of

all Product Backlog items completed during a Sprint and previous Sprints.

Through these iterative cycles, frequent feedback loops, and emphasis on collaboration, Scrum and other Agile methods (like Kanban, which focuses on visualizing workflow and limiting work in progress) aim to deliver value to the customer more quickly, reduce project risk, and effectively manage the inherent unpredictability often found in software development (Cohn, 2010).

ENSURING RELIABILITY: THE IMPERATIVE OF TESTING

Developing functional code and managing the process effectively are essential parts of programming, but they do not inherently guarantee a reliable or correct outcome. Software testing is therefore an indispensable activity within the development lifecycle. It involves the systematic execution of the software with the intention of finding defects (often called bugs) and assessing the overall quality of the product (Myers, Sandler and Badgett, 2011). Testing serves as a crucial quality assurance mechanism, providing confidence that the software behaves as expected and is fit for its intended purpose.

Why Software Testing is Non-Negotiable

The fundamental reason for testing is to identify errors before the software is delivered to end-users. Defects discovered after release can lead to significant problems, including financial loss, reputational damage, or system failures. While it is generally accepted that testing cannot prove a program is entirely free of errors (as testing all possible inputs and scenarios is usually infeasible for complex systems), it serves to reveal the presence of defects and provides critical information about the software's quality and the potential risks associated with its use (Kaner, Bach and Pettichord, 2002). Beyond finding bugs, testing helps verify that the software meets the specified requirements and can prevent previously fixed errors from reappearing (regression testing). It is important to differentiate testing from debugging: testing is the process of finding failures, whereas debugging is the subsequent process of locating and fixing the underlying faults that caused those failures (Zeller, 2009). Investing in thorough testing during development is generally far more cost-effective than addressing problems discovered later by users.

A Hierarchy of Tests: From Units to Systems

Software testing is typically conducted at multiple levels, starting with small components and progressing towards the entire system. This layered approach allows for methodical verification as the software is constructed. The main levels are:

- **Unit Testing:** This level focuses on testing the smallest testable parts of the software in isolation, such as individual functions, methods within a class, or modules. Unit tests are usually written and executed by the developers themselves during the coding phase. Their purpose is to validate that each component works correctly according to its specific design before it is integrated with others (Hunt and Thomas, 2002).
- **Integration Testing:** After individual units are tested, integration testing verifies the interfaces and interactions between them when they are combined. The goal is to detect issues that arise from the way components communicate or share data. For instance, does data passed from one module to another arrive correctly and get processed as expected? (Jorgensen, 2013).
- **System Testing:** Once the various components are integrated, system testing evaluates the complete system as a whole. This testing is performed against the overall

system requirements, both functional (does it do what it's supposed to do?) and non-functional (e.g., how well does it perform? Is it secure? Is it usable?). System testing provides an end-to-end perspective and is often carried out by an independent testing team to ensure unbiased evaluation (Pressman, 2005).

These levels often build upon each other, ensuring that basic components work before testing their interactions, and that interactions work before testing the entire system's behaviour (Sommerville, 2011).

Confirming Fitness for Purpose: Acceptance Testing

The final level of testing before software is typically released or deployed is Acceptance Testing. While system testing focuses on verifying the system against *technical specifications*, acceptance testing focuses on validating the system against *user needs* and *business requirements*. It essentially answers the question: "Does this software meet the needs of the user and the business, and is it acceptable for use?".

This type of testing is often carried out by the end-users or clients themselves, sometimes referred to as User Acceptance Testing (UAT), or by business analysts representing the stakeholders. The criteria for passing acceptance testing are usually defined based on the business goals and user requirements identified at the beginning of the project. Successful acceptance testing gives the stakeholders confidence that the developed system delivers the intended value and is ready for operational deployment (Black, 2016). This directly corresponds to the goal of Validation – ensuring the *right* system was built.

ARE WE BUILDING IT RIGHT? AND ARE WE BUILDING THE RIGHT THING?

The journey of software development, encompassing design, coding, management, and testing, ultimately aims to produce high-quality software that serves its intended purpose. Assessing whether this goal has been achieved requires answering two fundamental questions that were introduced earlier: "Are we building the system right?" and "Are we building the *right* system?". These questions encapsulate the crucial concepts of Verification and Validation, which are cornerstones of software quality assurance, heavily emphasized by influential figures like Boehm (1984). Understanding and addressing both V&V is vital for the success of any software project.

Introducing Verification and Validation

In software engineering, Verification and Validation (V&V) refer to distinct but related sets of activities performed throughout the development lifecycle to ensure software quality. Standard definitions help clarify their specific focus (based on IEEE Computer Society, 1990):

- **Verification** is the process of evaluating a system or component to determine whether the products of a given development phase satisfy the conditions imposed at the start of that phase. It essentially asks: "Are we building the system right?" according to the specifications?
- **Validation** is the process of evaluating a system or component during or at the end of the development process to determine whether it satisfies specified user requirements and needs. It essentially asks: "Are we building the *right* system?" for the user's context?

These are not just final checks but represent ongoing efforts to prevent errors and ensure the final product is both technically sound and functionally appropriate.

Verification: Meeting the Specifications

Verification is the process of confirming that the software meets its specified requirements and conforms to its design documentation. It focuses on checking the internal consistency and correctness of the software artifacts produced during each stage of development (Sommerville, 2011). Essentially, verification activities check if the product is being built according to the plan. A range of techniques are employed for verification. Formal or informal reviews and inspections allow human examination of requirements documents, design models, code, and test plans to identify defects and deviations from standards. Static analysis, often using automated tools, examines code without executing it to check for potential errors, security vulnerabilities, or violations of coding standards. Furthermore, much of the testing discussed previously serves verification aims: unit testing confirms individual modules meet their design, integration testing checks component interactions against interface specifications, and system testing verifies the complete system against documented requirements. Verification ensures that the software is built correctly as per the defined blueprints.

Validation: Fulfilling the User's Need

Validation, on the other hand, is the process of ensuring that the software, as built, satisfies the user's actual needs and intended use in their operational environment (Pressman, 2005). It checks whether the *right* problem is being solved and whether the software is fit for its purpose from the user's perspective. This involves assessing if the specifications themselves accurately captured the user's requirements. Common approaches to validation often emphasize user involvement. For example, prototyping early versions or mock-ups allows for crucial user feedback to ensure the development is heading in the right direction. Usability testing, where users are observed interacting with the system, assesses ease of use and task support. Finally, User Acceptance Testing (UAT), as highlighted earlier, allows users to determine if the system meets their needs sufficiently for deployment. It is entirely possible for software to be perfectly verified (i.e., built exactly according to specifications) but fail validation because the specifications did not accurately reflect the real user needs or context (Boehm, 1984).

How Process and Testing Serve V&V

Effective software development processes integrate both Verification and Validation activities throughout the lifecycle. Modern methodologies like Agile inherently support this integration. The focus within Agile on iterative development, frequent demonstrations of working software, and close collaboration with customers provides regular opportunities for ongoing Validation, helping to ensure the project remains aligned with true user needs (Cohn, 2010). At the same time, Agile practices such as having a clear "Definition of Done" for work items, which typically includes specific testing criteria, directly contribute to Verification. The different levels of testing also map naturally to these concepts. Unit, integration, and system testing predominantly focus on Verification, confirming the software's conformance to its technical design and specifications. In contrast, acceptance testing is the primary mechanism for Validation, confirming the software's fitness for its intended purpose and users (Black, 2016). By employing a combination of reviews, analysis, and testing techniques at various stages, and by using development methodologies that encourage feedback and adaptation, software projects can strive to answer both crucial questions affirmatively: ensuring the system is built right, and that the right system is built.

CONCLUSIONS

As explored throughout this paper, the act of programming transcends the simple translation of ideas into computer code. It represents a multifaceted discipline encompassing systematic processes, structured management, rigorous quality assurance, and a clear focus on achieving specific goals. Understanding programming in this holistic sense is crucial for appreciating its challenges and potential, particularly as computational methods become more integrated into diverse fields.

The journey from concept to functional software involves distinct stages, beginning with careful problem analysis and algorithm design before any implementation occurs. Whether using textual or graphical languages, the coding process itself benefits from established practices promoting clarity and maintainability, while debugging remains an inherent part of refinement. To manage the complexities of software creation, especially in collaborative or evolving projects, methodologies provide essential structure. While traditional models like Waterfall offer sequential control, the iterative and adaptive principles of Agile development have become highly influential, enabling teams to better navigate uncertainty and respond to changing requirements.

Furthermore, the reliability and trustworthiness of software depend critically on thorough testing. This is not a single event but a multi-layered process, involving unit, integration, system, and acceptance testing, each serving to uncover defects and build confidence at different levels of granularity. Underlying all these activities are the fundamental objectives of Verification and Validation, concepts crucial to software quality (Boehm, 1984). Verification ensures the software is built correctly according to its specifications, while Validation confirms that the software genuinely meets the user's needs and intended purpose. Successfully addressing both – building the system right, and building the right system – is the hallmark of effective software development, requiring conscious effort throughout the lifecycle.

Ultimately, programming is most effective when these elements – methodical process, appropriate management strategies, comprehensive testing, and a constant focus on both verification and validation – are seamlessly integrated. As algorithmic thinking and computational tools continue to permeate domains such as architectural planning (Terzidis, 2006), recognizing the full scope of the act of programming becomes increasingly important. Professionals in these fields who engage with creating or utilizing programmed tools benefit from understanding that reliable and valuable software arises not just from clever coding, but from a disciplined, quality-focused, and goal-oriented development process.

REFERENCES

Bach, J., Pettichord, B., Kaner, C., 2002. Lessons learned in software testing: a context-driven approach. John Wiley & Sons.

Beck, K., Beedle, M., Van Bennekum, A., Cockburn, A., Cunningham, W., Fowler, M., Grenning, J., Highsmith, J., Hunt, A., Jeffries, R., 2001. Manifesto for agile software development.

Black, R., 2016. Pragmatic software testing: Becoming an effective and efficient test professional. John Wiley & Sons.

Boehm, B.W., 1984. Verifying and Validating Software Requirements and Design Specifications. IEEE Software 1, 75–88. https://doi.org/10.1109/MS.1984.233702

Burnett, M.M., McIntyre, D.W., 1995. Visual programming. COmputer-Los Alamitos- 28, 14–14.

Cohn, M., 2010. Succeeding with agile: software development using Scrum. Pearson Education.

Fowler, M., Highsmith, J., 2001. The agile manifesto. Software development 9, 28–35.

Jorgensen, P.C., 2013. Software Testing: A Craftsman's Approach, Third Edition, 3rd ed. Auerbach Publications, New York. https://doi.org/10.1201/9781439889503

Knuth, D.E., 1997. The art of computer programming. Pearson Education.

McConnell, S., 2004. Code Complete: A Practical Handbook of Software Construction, 2nd ed. Microsoft Press, Redmond, Wash.

Myers, G.J., Sandler, C., Badgett, T., 2011. The art of software testing. John Wiley & Sons.

Pratt, T.W., Zelkowitz, M.V., Gopal, T.V., 1984. Programming languages: design and implementation. Prentice-Hall Englewood Cliffs, NJ.

Pressman, R.S., 2005. Software Engineering: A Practitioner's Approach. Palgrave Macmillan.

Royce, W.W., 1987. Managing the development of large software systems: concepts and techniques, in: Proceedings of the 9th International Conference on Software Engineering. pp. 328–338.

Schwaber, K., Sutherland, J., 2011. The scrum guide. Scrum Alliance 21, 1–38.

Sommerville, I., 2011. Software Engineering, 9/E. Pearson Education India.

Tedeschi, A., 2014. AAD Algorithms-Aided Design: Parametric Strategies using Grasshopper, 14th edition. ed. Le Penseur, Brienza.

Terzidis, K., 2006. Algorithmic Architecture. Routledge, London. https://doi.org/10.4324/9780080461298

Thomas, D., Hunt, A., 2002. Mock objects. IEEE Software 19, 22–24. https://doi.org/10.1109/MS.2002.1003449

Zeller, A., 2009. Why programs fail: a guide to systematic debugging. Morgan Kaufmann.

Subdivision Surfaces in Architecture

A Review of Techniques and Applications of Subdivision Surfaces in Architectural Design

Dorina Jakupi
TU Wien, Vienna, Austria
e11944045@student.tuwien.ac.at

This paper documents the current state of subdivision surface modelling in architectural design, examining its foundations, techniques and emerging applications. It reviews the most widely used subdivision schemes such as Catmull-Clark, Loop and Doo-Sabin, and analyzes their behaviour in relation to smoothness, and control topology. The paper also evaluates how subdivision surfaces are implemented in architectural workflows, particularly in parametric modelling environments. The aim is to clarify the role of subdivision surfaces in contemporary architecture and its help in planning.

Keywords: *Subdivision surface modelling, Architectural geometry, Computational design.*

INTRODUCTION

Subdivision Surfaces is a modelling method that creates smooth and continuous shapes from simple geometry. The process begins with a coarse polygon mesh to which a set of recursive rules is applied in order to refine its shape, which then gradually increases its smoothness at each step. This makes it possible to design curved and complex forms without needing to build detailed geometry from the start. This technique first became popular in computer animation, where smooth and flexible surfaces are essential. In the recent years it has started to appear more in architectural design, architects and designers use subdivision surfaces to explore form, create fluid geometries and produce models that can be connected directly to digital fabrication.

The method works well with other tools like parametric design softwares and structural simulations. It also supports non-destructive workflows where geometry and performance are developed in parallel. Designers can modify the underlying mesh while maintaining control over the refined surface. This flexibility makes them especially useful in early design phases where changes are frequent, as well as in later stages where precision and control are cruical.

As architectural projects demand both geometric complexity and technical precision, subdivision surfaces can provide a way to combine the creative modelling and the practical application. This growing relevance makes it important to understand how these techniques function, where they are being used and how they compare to other surface modelling approaches.

Background and Core Principles

The process of subdivision surfaces works by applying a set of rules that split and average the geometrical faces, edges and vertices of the given geometry. With each step, the mesh becomes smoother and moves closer to a curved surface. This iterative process produces what is known as the limit surface*, a continuous form that emerges from a relatively simple base mesh. This approach was first developed in computer graphics to handle complex shapes more flexibly than the traditional polygon and NURBS modelling.

There are multiple subdivision methods that are used in design today.

Catmull-Clark

Developed in 1978 by Edwin Catmull and Jim Clark, the method generalizes bicubic B-Spline surfaces to meshes of arbitrary topology, which makes possible the creation of complex continuous surfaces from irregular base geometries (Catmull and Clark 1978). It is one of the most widely adopted methods of subdivision and it works best with quad meshes.

The Catmull-Clark scheme works by repeatedly subdividing the mesh through a series of defined steps which in each iteration compute new face points as the average of all vertices that define a face. Edge points are calculated as the average of the two adjacent face points and the two vertices that define the edge. Vertex points are then updated based on a weighted average involving adjacent face points, edge midpoints and the original vertex itself. This process is repeated recursively generating a limit surface that is smooth and continuous across the mesh, including at extraordinary vertices where the mesh deviates from regularity.

The weighting scheme used to calculate new vertex positions was selected by Catmull and Clark primarily for the quality of the resulting shapes rather than derived from a purely mathematical optimization. Nonetheless, their work includes a detailed mathematical demonstration showing that the method converges to surfaces equivalent to bicubic B-Splines.

Doo-Sabin

The Doo-Sabin subdivision algorithm was presented in 1978 by Daniel Doo and Malcolm Sabin as a method for generating smooth surfaces from polygonal meshes. It was developed alongside the Catmull-Clark method, however the Doo-Sabin subdivision method extends the principles of tensor product B-spline surfaces to meshes of arbitrary topology, enabling the generation of smooth forms without the strict constraints of rectangular control grids (Doo and Sabin, 1978).

In contrast to Catmull-Clark, which focuses on quad-dominant meshes and creates new points at face centers, edge midpoints and vertex positions, the Doo-Sabin algorithm produces new vertices for each face and each vertex in the original mesh. As a result, each original face gives rise to a new face with the same number of edges, and new faces are also created for each original edge and vertex. Over successive iterations, the mesh becomes more dense of quadrilateral faces and approaches a smooth limit surface.

A key feature of Doo-Sabin subdivision is to generate a regular, grid like mesh structure as the number of iterations increases. This feature is advantageous in architectural applications where uniform panels or consistent face orientation are required. Furthermore, it can be applied to a variety of base geometries, including those with irregular connectivity or non-quad elements, because it supports inconsistent input topologies.

The Doo-Sabin method shows a C^1 continuity at extraordinary points and a C^2 continuity in regular points, which creates smooth visual transitions across the surface. Unlike Catmull-Clark, which maintains a close connection to B-spline theory, Doo-Sabin directly applies affine combinations of vertex positions from the original mesh to generate each new vertex, making the underlying logic more geometrically intuitive in some implementations(Doo and Sabin, 1978).

Loop

The Loop subdivision scheme was introduced by Charles Loop in 1987, a subdivision method designed specifically for triangular meshes. It was developed as a generalization of quartic box splines and it smoothes the surfaces through a recursive process of vertex insertion and repositioning. Its algorithm makes the Loop method well equipped to deal with topologies that have triangle based mesh. (Loop, 1987).

In each iteration, the algorithm subdivides every triangle in the mesh into four smaller triangles. new vertices are introduced at the midpoint of existing edges, and the positions of both new and old vertices are recalculated using weighted averages based on the surrounding topology. The weights are carefully chosen to ensure that the resulting limit surface maintains the smoothness.

The Loop subdivision creates surfaces that are C^2 continuity throughout the entire surface, except at extraordinary vertices where the surface remains C^1 continuous.

One of the main advantages of all subdivision methods is how well they handle irregular topologies. Unlike NURBS, which often require complex patching and trimming, subdivision can maintain smoothness across vertices with any number of connections. This makes it much easier to model freeform geometry without breaking surface continuity and it is also effective for modelling geometries that include branching structures.

These principles make subdivision surfaces a valuable tool in architectural design, they support a balance between flexibility and precision.

Computational Methods and Performance

The main reason subdivision surfaces are used as a method is because it creates smooth and continuous surfaces even when the shapes are complex. Many subdivision methods keep the surface smooth not just where faces meet but also across sharp turns and branching areas, this being the key advantage over traditional polygon or NURBS modelling. Many of these schemes aim for at least C^1 continuity, meaning that the surface remains smooth across joins and irregular points. This is difficult to achieve with traditional NURBS surfaces, which often require manual patching to stay visually seamless. Zorin (2000) introduced a widely used method for analyzing the smoothness of subdivision surfaces across different mesh topologies, showing that schemes like Catmull-Clark and Loop can maintain reliable continuity even at extraordinary vertices. This ability to handle branching and non uniform mesh structures makes subdivision surfaces especially useful for modelling freeform architecture with high surface quality.

Another important idea is adaptive subdivision. This means the model doesn't need to be refined everywhere, only in the parts where more detail is needed. For example, curves, corners, or detailed areas can have more subdivisions, while flat or simple areas can stay coarse. This saves time and keeps the model lightweight. In early stages of design, using a simple version is helpful for fast testing. Later, more detail can be added only where it matters. Adaptive subdivision lets designers balance speed and detail depending on the task. Amresh (2003)

Applications in Architectural Design

Subdivision surfaces have quickly gained an important place in the early stages of architectural design, particularly in the creation of conceptual models. Their ability to create fluid, continuous geometry enables designers to test spatial ideas without being constrained by rigid constraints. This allows for a more exploratory designing process in which objects can be manipulated in an intuitive and straightforward way. This flexibility is proven to be very helpful in the early stages of architectural competition based work, where originality and visual coherence is usually valued more than technical accuracy. This help mostly comes by its ease to incorporate into workflows where quick iterations and spatial exploration are crucial by designers due to the ease with which it may be used to simulate intricate branching systems or irregular volumes.

Those workflows can also be parametric environments, subdivision surfaces help their modelling strategies by offering a geometric layer that changes in response to the change in the parameters. These parameters can be embedded within the controls of the mesh that was altered by the subdivision, which in turn updates accordingly to the variable inputs. This integration enables the creation of responsible design systems that combine geometric continuity with algorithm control. An example would be the workflow available in Grasshopper (Rhino 3D Plugin), when embedded within, subdivision techniques support workflows where form generation, environmental feedback and user defined constraints can operate simultaneously.

Subdivision surface as a concept can also be helpful not in just planning design concepts of structures but also in terrain work and pathways that lead to structures or to the creation of structures from such pathways. This gives planners a canvas for thinking about the land as a continuous field instead of a patchwork of individual detached cells. By turning slopes or noise levels into a web of control points, routing algorithms can trace paths that curve naturally around obstacles. Instead of snapping to a rigid grid, the planner can assign weight values to the mesh and let subdivision algorithms find routes that flow with the topography. This results in an organic circulation network that responds to the site's subtle shifts in elevation or land use.

One of the main challenges with using subdivision surfaces for planning is managing the balance between detail and performance. When large models are subdivided, a uniform refined mesh can become extremely heavy and slow down other tasks. Even with adaptive subdivision schemes that refine only where needed still require careful tuning of error thresholds and curvature metrics, or else planners end up trading off interactivity for precision. In practice this signifies that planners always have to keep in mind the performance of the model and simplify their control nets or limit the extent of subdivision in order to keep it responsive, which can undermine the very purpose, smoothness and continuity that they were seeking.

REFERENCES

Zorin, D. (2000). A Method for Analysis of C^1 -Continuity of Subdivision Surfaces. Computer Aided Geometric Design. Available at: https://cims.nyu.edu/gcl/papers/zorin2000mac.pdf

Amresh, A., Farin, G., Razdan, A. (2003). Adaptive Subdivision Schemes for Triangular Meshes. in Farin, G., Hamann, B., Hagen, H. (eds) Hierarchical and Geometrical Methods in Scientific Visualization. Mathematics and Visualization. Springer, Berlin, Heidelberg. Available at: https://doi.org/10.1007/978-3-642-55787-3_19

Catmull, E., Clark, J., (1978) *Recursively generated B-spline surfaces on arbitrary topological meshes*. Computer-Aided Design. Available at: https://doi.org/10.1016/0010-4485(78)90110-0

Doo, D., & Sabin, M. (1978). *Behaviour of recursive division surfaces near extraordinary points*. Available at: https://doi.org/10.1016/0010-4485(78)90111-2

Loop, C. (1987). *Smooth subdivision surfaces based on triangles*.

Intelligent kinetic façades

The algorithmic thinking behind the movement

[1]Raphael Auffarth,
[1]Technical University of Vienna
[1]e11702811@student.tuwien.ac.at

The concept of kinetic facades in literature as well as in architecture is not a new thing. The implementation of driving them by an algorithm to improve the energy efficiency of a building on the other hand is a relatively new and interesting aspect of kinetic facades. This Paper will give basic information about the state-of-the-art projects in architecture as in literature. It will explain the key concepts of kinetic facades and the algorithmic thinking behind the movement.

Keywords: *kinetic, facades, algorithmic design, algorithmic thinking.*

INTRODUCTION

The first exploration of kinetic movement and kinetic facades in architecture have been done in the early 20th century. In 1919 to 1920 the Russian artist Vladimir Tatlin imagined a dynamic structure that would change over lifetime, the Monument to the Third International. In 1933 Yakov Chernikhov published his book "101 Architectural Fantasies" giving theoretical examples that were thought to be integrated into to architecture. Later, architects started experimenting and implementing moving parts in their architectural projects. Buckminster Fuller for instance used moving parts in his designs for the Geodesic domes. In 1970 William Zuk published his book "Kinetic Architecture" which led to more practical implementation in architecture such as Buckminster Fullers concept of tensegrity. From then on kinetic architecture was a recurring aspect in architecture. Since then, technology has improved more and more, especially technologies like robotic or parametric design and 3D printing and the interest in actively controlled movements has increased.

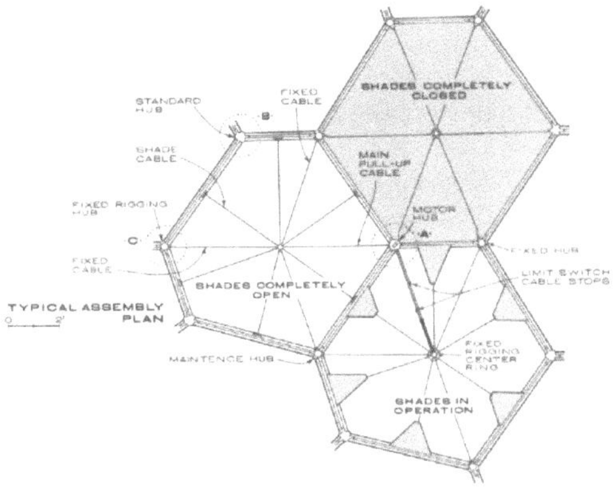

Figure 1: Detail of the US Pavilion shading system. (Kalin,Survey of Building Materials, Systems, and Techniques.)

DEFINING KEY CONCEPTS

Kinetic architecture and more general kinetic movement can be classified in several groups as sliding, rotating, opening-closing and folding. (Yeşim İLERİSOY and Pekdemir Başeğmez, no date) To be able to perform this movement, movable architecture must consist of structure, connections, actuators, materials and control systems. (Osama, 2014) On the other hand kinetic movement can be categorized by its purpose of use of motion. These are intelligent architecture, interactive architecture, responsive architecture, adaptive architecture, and responsive kinetic architecture. (Elmokadem *et al.*, 2016) In the following paragraphs the focus is on kinetic facades and the algorithms which they are controlled by.

What are kinetic facades?

Without harming the overall structural integrity, kinetic facades describe the actual movement or motions through geometric transformation in space that affect the changing state or material properties or physical structure of the building facades. Kinetic facades can be used to improve aesthetic qualities, respond to changing environmental conditions, and carry out tasks that static facades would not be able to. (Sharaidin BArch, no date)

What are responsive facades?

Based on how they affect the building, responsive facades fall into four major categories: noise, humidity, natural air current, and solar radiation.

Since each of these factors affects how comfortable people are within a structure, it can be helpful to adapt the façade to the external conditions. (*Adaptive Façade: Concept, Applications, Research Questions | Enhanced Reader*, no date) By integrating such technology responsive facades can save up to 50% of the energy needed for heating, ventilation and air conditioning (HVAC). (Kadhim, Al Maamory and Al Ghaban, 2025)

How is it controlled?

In order to move a kinetic façade it has to be controlled either directly, meaning by remote, or indirectly by a computer. For indirectly controlled facades there are commonly two different approaches either extrinsic, an external decision-making unit or an intrinsic control, an internal self-adjusting system. Both approaches need data which is processed, this data can come from sensors, internal backup and restored information, manually entered information by users, and by being connected online. (Heidari Matin and Eydgahi, 2022)

ACTIVE CONCEPTS AND EXAMPLES

Examples which are using active responsive façade elements utilizing the various concepts mentioned in the previous section are reviewed in the paragraphs to follow.

Light Control kinetic facade systems

- Institut du Monde Arabe, Jean Nouvel, 1987

In 1987 French architect Jean Nouvel designed the Institut du Monde Arabe located in Paris, France. Its façade uses photosensitive motor controlled shutters made out of aluminum that open and close automatically to regulate the sunlight and the resulting heat that is entering the building. The movement is controlled by a central controller that steers onwards hydraulic actuators. (*Institut du monde arabe (IMA) — Ateliers Jean Nouvel*, no date)

Figure 2 detail of the shutters (photo: Jakob Schoof)

- Al Bahar Towers, Aedas Architects, 2012

The London based architecture office Aedas Architects built the in 2012 finished project Al Bahar Towers. It uses 2099 translucent PTFE fiberglass panels. Each panel consists of three parts which open or close themselves automatically following the course of the sun using hydraulic actuators. The mechanism is controlled centrally and uses electro-mechanical technology together with light sensors. Using this technology the architects saved 50% in energy consumption in comparison to conventional buildings. (*Al Bahar Towers, Abu Dhabi - Architekturobjekte - heinze.de*, no date)

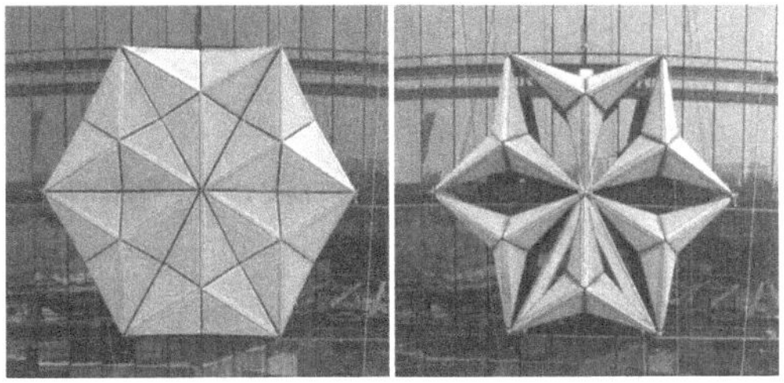

Figure 3: Dynamic, sunlight-responding shading screen for Al Bahar Towers, reproduced from aedas.com/en

Wind controlled kinetic Façade

- Greenway Self-Park, Hok, 2010

The 11-story parking garage, designd by HOK architect, follows another interesting strategy of kinetic facades. It uses vertical wind turbines located on the southwestern corner of the building to electrify the power of wind. With the energy gained the building can provide enough energy to run the exterior light of the building as well as returning power back to the city grid over the year.(*Greenway Self-Park by HOK - Architizer*, no date)

Figure 4 : Wind turbines at Greenway Self-Park

PASSIVE CONCEPTS AND EXAMPLES

Additional concepts for kinetic facades which are not computer-driven but are nonetheless equally interesting are presented in the sections to follow.

Wind Responsive Kinetic Facade System

- Wind Veil, Ned Kahn Studios, 2000

In Charlotte, North Carolina, USA a 6-story parking garage is cover with 80000 aluminum plates that can freely move with the wind. The kinetic façade is designed by Ned Kahn and has no control system and the only actuator is the wind. There for it is a passive responsive façade which not only affects the aesthetics of the building façade but furthermore it provides natural ventilation as well as shading in the inside of the garage. (*Wind Veil – NED KAHN*, no date)

Material-Based technology responsive facades

- Bloom, DOSU Studio, 2011

Inspired by Victorian-era underwear, "Bloom" is a sun-tracking structure designed by Doris Sung in

2011 that combines structural innovation, material experimentation, and computational design to create form that responds to climate change. As temperatures rises, its sheets out of thermobimetal curl without the need for external energy. Through this effect it reduces the need for air conditioning in buildings. These dynamic materials have a big impact on facade design and urban aesthetics.

CONCLUSION

Kinetic façades represent a dynamic intersection of architecture, engineering, and computational design, offering innovative solutions for energy efficiency, environmental responsiveness, and aesthetic expression. From early conceptual visions to contemporary implementations using algorithmic control and responsive materials, the evolution of kinetic façades highlights the growing importance of adaptability in architectural design. The reviewed examples demonstrate a range of strategies, from active systems powered by sensors and actuators to passive approaches relying on materials or environmental forces.

To fully realize the potential of kinetic façades, it is crucial not only to refine existing strategies but also to combine them in creative ways and continue developing new ones. By integrating multiple approaches—mechanical, material-based, algorithmic, and beyond—designers can create façades that are more intelligent, efficient, and contextually responsive. Continued interdisciplinary exploration will be key in pushing the boundaries of what kinetic architecture can achieve in the future.

REFERENCES

Adaptive Façade: Concept, Applications, Research Questions | Enhanced Reader (no date).

Al Bahar Towers, Abu Dhabi - Architekturobjekte - heinze.de (no date). Available at: https://www.heinze.de/architekturobjekt/al-bahar-towers-abu-dhabi/12556305/ (Accessed: 8 June 2025).

Elmokadem, A. *et al.* (2016) 'Volume 7 Issue 4, April 2018 www.ijsr.net Licensed Under Creative Commons Attribution CC BY Kinetic Architecture: Concepts, History and Applications', *International Journal of Science and Research (IJSR) ISSN*, pp. 7–296. Available at: https://doi.org/10.21275/ART20181560.

Greenway Self-Park by HOK - Architizer (no date). Available at: https://architizer.com/projects/greenway-self-park/ (Accessed: 8 June 2025).

Heidari Matin, N. and Eydgahi, A. (2022) 'Technologies used in responsive facade systems: a comparative study', *Intelligent Buildings International*, 14(1), pp. 54–73. Available at: https://doi.org/10.1080/17508975.2019.1577213.

Institut du monde arabe (IMA) — Ateliers Jean Nouvel (no date). Available at: https://www.jeannouvel.com/projets/institut-du-monde-arabe-ima/ (Accessed: 8 June 2025).

Kadhim, B.M., Al Maamory, A.S. and Al Ghaban, A.M. (2025) 'Advancing adaptive facades for a sustainable future in building design', *Edelweiss Applied Science and Technology*, 9(1), pp. 801–845. Available at: https://doi.org/10.55214/25768484.V9I1.4251.

Osama, Y. (2014) 'INTERACTIVE MOVEMENT IN KINETIC ARCHITECTURE', 42(3), pp. 816–845.

Sharaidin BArch, K. (no date) *Kinetic Facades: Towards design for Environmental Performance.*

Wind Veil – NED KAHN (no date). Available at: https://nedkahn.com/portfolio/wind-veil (Accessed: 8 June 2025).

Yeşim İLERİSOY, Z. and Pekdemir Başeğmez, M. (no date) 'Conceptual Research of Movement in Kinetic Architecture'. Available at: http://dergipark.gov.tr/gujs (Accessed: 8 June 2025).

Comparative Analysis of Two Air Change Rate Models for Single-Sided Natural Ventilation

Hanna Weltken
TU Wien
e12331435@student.tuwien.ac.at

Natural ventilation is an increasingly relevant strategy in reducing building energy demand and improving indoor environmental quality. This paper compares two fundamental methods for estimating air change rates in single-sided natural ventilation: a Bernoulli-based model describing bidirectional flow through large openings and a simplified approach introduced by Warren and Parkins. Both models form the theoretical foundation of several national ventilation standards. The comparison highlights key differences in complexity, accuracy and applicability. While the Bernoulli-based formulation offers high physical accuracy, it is computationally intensive. In contrast, the Warren and Parkins model provides a simplified and practical alternative suitable for early-stage planning.

Keywords: *Natural Ventilation, Air Change Rate, Single-sided Ventilation*

INTRODUCTION

The demand for indoor thermal comfort and air quality results in high energy consumption by heating, ventilation and air conditioning (HVAC) systems. It is estimated that buildings account for approximately 40 % of total global energy consumption and over 30 % of CO_2 emissions [1]. Since HVAC systems are responsible for around 50–60 % of a building's total energy use [2], natural ventilation is gaining importance as an energy-efficient strategy to ensure indoor comfort.

To effectively apply natural ventilation in practice, it is essential to understand and accurately estimate the air change rate (ACR). The ACR is a key input in building simulation and planning, as it directly affects energy demand, thermal comfort and indoor air quality assessments.

This paper focuses on two widely used methods for estimating air change rates in single-sided natural ventilation scenarios. The Bernoulli-based ideal distribution and the simplified wind and stack approach by Warren and Parkins. Both models form the theoretical basis for most current ventilation standards, including ÖNORM B 8110-3 (Austria), VDI 2078 (Germany) and ASHRAE (United States). These standards build upon the same physical principles but differ in their assumptions and simplifications. Understanding these foundational models makes it easier to interpret national standards and evaluate where and how they can be applied effectively.

FUNDAMENTALS OF NATURAL VENTILATION

Natural ventilation in buildings can generally be classified into two main strategies: cross ventilation and single-sided ventilation. In cross ventilation openings are positioned on opposing sides, allowing air to flow directly through the space. In contrast single-sided ventilation relies on openings located on only one side of the building, with both the inflow and outflow of air through the same opening.

Despite its higher efficiency, cross ventilation is often difficult to implement in practice. Factors such as compact building layouts, fire safety requirements, security constraints and privacy concerns frequently limit the possibility of continuous airflow across rooms [3]. As a result, single-sided ventilation remains an common solution.

Air movement in single-sided natural ventilation is primarily driven by three physical mechanisms: buoyancy forces, wind pressure and their combined interaction [4]. Buoyancy-driven flow occurs due to temperature differences between indoor and outdoor environments, which lead to pressure differences that drive air through the openings.

In contrast, wind-driven ventilation is significantly more complex. Wind generates direct pressure differences along the building envelope, but its speed and direction are highly variable and difficult to predict. These unsteady characteristics of wind result in turbulent and non-linear airflow patterns both outside and within the building [5].

In most real-world situations, buoyancy and wind forces act simultaneously. The resulting airflow depends on the interaction of these forces. Depending on whether the pressure differences reinforce or counteract each other, the overall ventilation effect may be increased or diminished [6].

The Orifice Equation as the Basis of Many Models

Many calculation approaches used to estimate the air change rate through window openings are based on the orifice equation. It is a physical model that describes the volumetric airflow rate through an opening under a pressure difference:

$$Q = C_d A \sqrt{2gh} \tag{1}$$

In this equation Q represents the volumetric airflow rate, A is the effective area of the opening, C_d is the discharge coefficient, g denotes gravitational acceleration and h is the pressure-equivalent height difference.

The equation is derived from Bernoulli's theory for incompressible flow and describes the conversion of potential energy (pressure differences) into kinetic energy (air velocity). It forms the foundation for many standardized and empirically adapted methods used to assess natural ventilation performance [7].

Flow Through Large Openings – Bernoulli-Based Ideal Distribution

A specific case of natural ventilation through a single large opening involves bidirectional airflow, in which cooler outside air enters through the upper part of the opening while warmer indoor air exits through the lower part. This scenario typically occurs under purely buoyancy-driven conditions, so without wind influence.

Between the inflow and outflow zones lies the neutral pressure level h, where the pressure difference across the opening is zero. The airflow along the vertical axis of the opening follows an idealized velocity profile that can be derived from the Bernoulli equation. The resulting horizontal air velocity $V(y)$ at a given height y is described by:

$$V(y) = -C_d(y-h)\sqrt{2g\frac{\Delta p}{p_r}(y-h)} \tag{2}$$

C_d is the discharge coefficient accounting for contraction and friction losses and g is the gravitational acceleration. The term Δp denotes the air density difference between the outdoor and indoor environment and p_r represents the reference air density, which is typically taken as the outdoor density The vertical distance from the neutral level is represented by $y - h$.

The flow pattern is illustrated in Figure 1, which shows the typical bidirectional distribution of airflow velocities along a single opening under buoyancy-driven conditions.

The equation assumes parallel streamlines and no mixing between the incoming and outgoing air. While it allows for a theoretically accurate description of the velocity distribution across the opening, it is rarely used directly in practical applications due to its complexity in calculation The numerical demand is high, because a separate velocity value must be calculated at each vertical position along the opening [8].

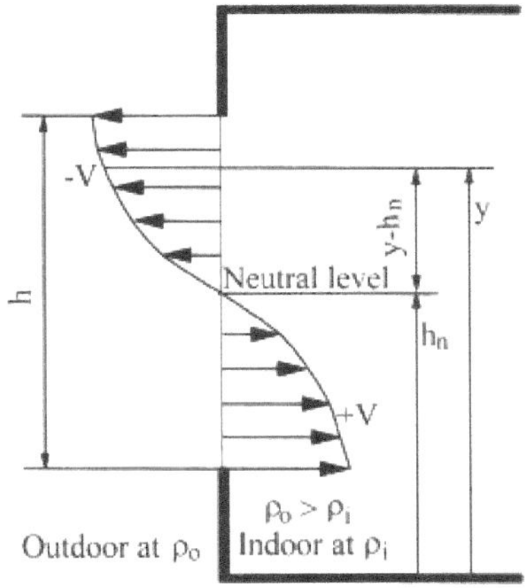

Figure 1 Velocity Distribution at single sided opening [8]

Warren and Parkins – Combined Wind and Stack Model

Warren and Parkins developed another, simpler model to estimate air change rates through window openings under single-sided natural ventilation. Their approach is also based on the orifice equation, but the distribution of incoming air is simply modelled as a triangle. It treats the effects of temperature differences (stack effect) and wind pressure separately [9].

The buoyancy-driven component of the airflow rate is calculated using the following expression:

$$Q_{Stack} = \frac{1}{3} A C_d \sqrt{\frac{\Delta T H g}{T_{av}}} \qquad (3)$$

Here, the notation for parameters such as A, C_d and g is consistent with the general orifice formulation introduced earlier. In addition ΔT represents the temperature difference between the indoor and outdoor air, H is the vertical height of the opening and T_{av} is the average air temperature in Kelvin.

The wind-driven component is calculated using a simple linear relationship:

$$Q_{Wind} = 0.0025 A v_{wind} \qquad (4)$$

where v_{wind} is the wind speed at the height of the opening.

Rather than combining the stack and wind effects, the model assumes that the larger of the two values represents the total ventilation rate:

$$Q = max\,(Q_{Stack}, Q_{Wind}) \qquad (5)$$

This approach enables a fast and practical estimation of ventilation performance with minimal input data, making it well-suited for early-stage design analysis or comparative studies in naturally ventilated buildings.

COMPARISON OF THE TWO METHODS

The Bernoulli-based formulation offers a detailed and theoretically accurate representation of the velocity distribution along a large opening. However, it requires complex calculations, as the air velocity must be determined individually at each vertical position. This level of detail allows for precise modeling but also results in high computational effort, which makes the method primarily applicable in CFD simulations or academic validation studies rather than in routine design practice.

In contrast the Warren and Parkins model simplifies the flow distribution by assuming a triangular profile across the opening. Instead of calculating velocity continuously along the height, the method uses the centroid of the triangular area to represent the average flow rate. This corresponds to taking one-third of the height as the effective point of flow, which explains the appearance of the 1/3 factor in the stack-driven flow equation. This simplification enables fast estimations with minimal input and is more practical for early-stage design, parametric studies and integration into simulation tools.

Because of its simplicity and ease of implementation, the Warren and Parkins model is used as the basis for several national ventilation standards. Its assumptions and structure are reflected in ÖNORM B 8110-3 (Austria), VDI 2078 (Germany), and the ASHRAE Handbook (United States), each with country-specific modifications.

CONCLUSION

Both models offer distinct advantages depending on the application. The Bernoulli-based model provides high physical accuracy and is useful for detailed theoretical investigations or model validation. The Warren and Parkins method enables fast and accessible estimation of ventilation rates, making it well-suited for practical use and integration into national standards. However, its simplifying assumptions result in a loss of precision and can lead to under- or overestimation of real ventilation behavior. A critical understanding of each model's assumptions, strengths and limitations is therefore essential for selecting the appropriate method based on the design stage, simulation requirements and regulatory context.

REFERENCES

[1] Abergel, T. et al. (2017) *Energy Technology Perspectives 2017: Catalysing Energy Technology Transformations*. OECD.

[2] Perez-Lombard, L. and Ortiz, J. (2008) 'A review on buildings energy consumption information', *Energy and Buildings*, 40(3), pp. 394–398.

[3] Wang, J. and Wong, N.H. (2017) 'Assessment of single-sided natural ventilation driven by buoyancy forces through variable window configurations', *Energy and Buildings*.

[4] Awbi, H.B. (2003) *Ventilation of Buildings*. 2nd edn. New York: Spon Press.

[5] Ai, Z.T. and Mak, C.M. (2014) 'Analysis of fluctuating characteristics of wind-induced airflow through a single opening using LES modeling and the tracer gas technique', *Energy and Buildings*.

[6] Larsen, T.S. and Heiselberg, P. (2008) 'Single-sided natural ventilation driven by wind pressure and temperature difference', *Energy and Buildings*.

[7] Jiang, Z. and Kobayashi, T. (2022) 'Validity of Orifice equation and impact of building parameters on wind-induced natural ventilation rates', *Building and Environment*.

[8] Gan, G. (2000) 'Effective depth of fresh air distribution in rooms with single-sided natural ventilation', *Energy and Buildings*.

[9] Warren, P.R. and Parkins, L.M. (1985) 'Single-sided ventilation through open windows', in *Proc. Conf. Thermal Performance of the Exterior Envelopes of Buildings*. ASHRAE SP 49, pp. 209–228.

State-of-the-Art in Indoor Lighting Estimation: Emphasis on Light Octree

Besfort Ramadani[1]
[1]*TU Wien*
[1]*e12402555@student.tuwien.ac.at*

Estimating indoor lighting is central to applications in augmented reality (AR), virtual reality (VR), and photorealistic rendering. Recently, new methods have emerged that attempt to estimate lighting with spatial coherence from limited input data (for example, a single RGB image). This report investigates the "Light Octree" method, places it within the context of current research, and compares it with other important methods to show the current state-of-the-art.

Keywords: *Indoor Lighting Estimation, Voxel Octree, Differentiable Rendering, Augmented Reality, Neural Scene Representation, Spatially Coherent Illumination.*

INTRODUCTION

When placing virtual objects into real scenes in indoor environments, correct lighting estimation is key. When the virtual object is viewable from unknown viewpoints, traditional approaches tend to involve site visits which require reading a lot of reference material, gathering lots of data or expensive equipment, thus, limiting feasibility. The challenges in light estimation from fewer inputs include estimating lighting conditions that are complicated, maintaining spatial coherence, and ensuring computational performance.

TRADITIONAL APPROACHES

Indoor lighting estimation methods used to be based on global illumination models that assumed a uniform light condition across an indoor scene; for example, environment map estimations from light probes, chrome spheres, etc. Because these methods did not take account of the spatial variation of light, inserting realistic virtual objects into scenes was not very accurate.

EMERGENCE OF SPATIALLY VARYING LIGHTING ESTIMATION

To counter the drawbacks to global models, a set of spatially-varying lighting estimation approaches were developed to capture spatial lighting variations. Some examples include:

- Deep Parametric Indoor Lighting Estimate: This approach uses a set of discrete 3D lights that contain photometric and geometric parameters of a lighting instance, helping the system utilize more localized light estimates.
- DSGLight: A method that uses depth-augmented spherical Gaussians and graph convolutions to infer a lighting representation from a single low dynamic range image.
- StyleLight: A method that utilizes a paired dual-StyleGAN panorama synthesis network to render high-dynamic-range indoor panoramas from a single low field-of-view image.

Consequently, these methods extended the spatial accuracy of lighting estimation compared to global models, but typically at the expense of complicated models and/or high computational costs.

LIGHTOCTREE: A LIGHTWEIGHT SOLUTION

The LightOctree method provides a new solution by creating a lightweight approach to estimating indoor spatially coherent lighting from a single RGB image. Key points include:

- Voxel Octree Representation: LightOctree uses a voxel octree representation of the scene when estimating lighting, as light source presence in indoor environments is sparse. The octree structure serves as a hierarchical data structure and can also be used to store the scene lighting values.
- Differentiable Cone Tracing Rendering Layer: The differentiable cone tracing rendering layer allows the model to render scenes using both high-frequency and low-frequency lighting information and makes sure these individual components are accurately represented, see Figure 1.
- End-to-End Training: The framework incorporates depth estimation and lighting estimation in a single pipeline, so the entire approach can be trained and used in inference seamlessly.

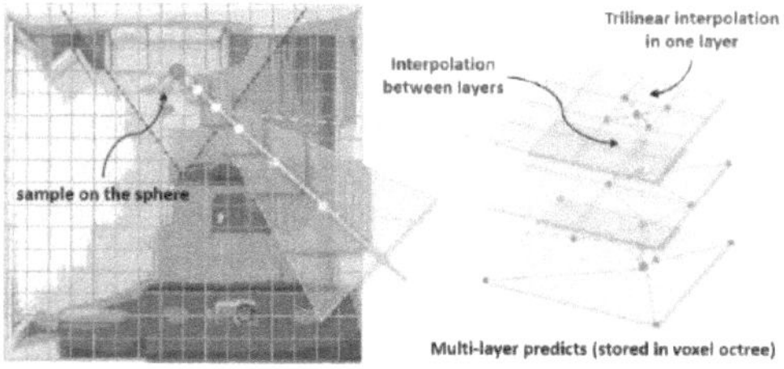

Figure 1 Differentiable rendering in Light Octree

COMPARATIVE ANALYSIS

When compared to its nearest state-of-the-art methods:

- Efficiency: The use of voxel octrees in LightOctree, for example, means that users will have less overhead computationally compared to some other methods that make it easier to run in a real-time application.
- Accuracy: While DSGLight and StyleLight yield very accurate results, they come at the expense of increased complexity and burden on users. LightOctree was evaluated with some complexity, however, it allowed for a compromise of accuracy and efficiency.
- Flexibility: The way in which the framework has been designed offers a very easy interface for a variety of real-time applications such as AR and image editing without the burden of excess preprocessing or the reliance on specialized equipment.

In comparison to previous methods, Figure 2 emphasizes how Light Octree delivers visually coherent renderings of virtual objects when lighting is only estimated and how it improves on both spatial consistency and photorealism. Light Octree comes out on top in rendering with more natural shading and smoother transitions as compared to previous methods in indoor scenes. These results demonstrate the effectiveness of integrating Light Octree into practice.

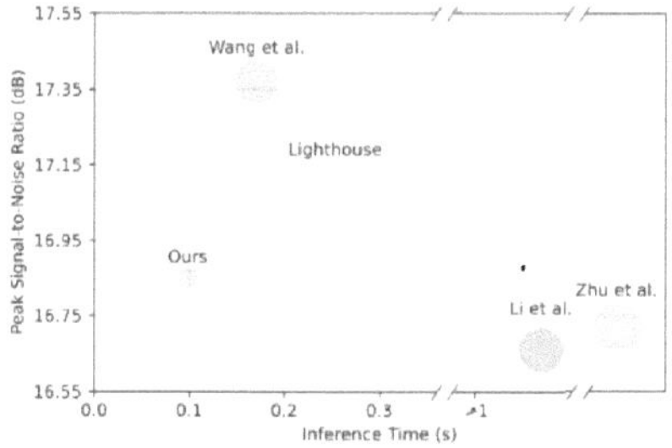

Figure 2 Comparison of Light Octree with other lighting estimation methods

APPLICATIONS AND FUTURE DIRECTIONS

LightOctree's simplicity and effective implementation allows it to be applied in many different contexts such as:

- Augmented and Virtual Reality: Using real-time estimates of the lighting will allow virtual objects to feel more realistic in AR and VR contexts.
- Image Editing: Providing a more accurate estimate of the lighting in a scene based on a photograph with useful lighting information will improve the certainty of integrating and editing elements in a photograph.
- Robotics and Autonomous Systems: Understanding the lighting conditions in which the autonomous, robotic or drone system will perform will help understand how the lighting will influence navigation and recognition of objects.

Future research directions may be considered:

- 1Integration with Generative Models: Light Octree can be used with generative adversarial networks (GAN) which could improve realism.
- Expansion to Outdoor Environments: Adapting this method of configuration for outdoors where natural lighting is a factor.
- User-Guided Editing: Providing users the opportunity to view changes to a surface by interactively editing the lighting estimates to determine visual and design capabilities.

EVALUATION AND RESULTS

The performance of LightOctree has been tested on synthetic indoor datasets like FutureHouse that are paired with ground-truth HDR environment maps and geometry. LightOctree provides strong spatial coherence and realism when relighting virtual objects, even under complicated lighting distributions.

Quantitative evaluations show that LightOctree can achieve similar (or better) accuracy compared to previous volumetric methods with far less computational costs. Its memory and FLOP efficiencies arise from its voxel octree structure, which is $O(n^2)$ scaling, compared the $O(n^3)$ complexity of dense voxel grids. Moreover, LightOctree has relatively good rendering quality with respect to camera pose ($O(n)$), making it beneficial in dynamic AR/VR contexts.

Qualitative comparisons (see Figure 2) show improved shadow consistency, and better surface shading across the rest of the virtual object. This improvement across the whole virtual lighting subject can be seen when compared with other methods, like DSGLight and StyleLight. Especially when multiple or localized light sources were present, LightOctree allows for smooth transitions and accurate occluded light across the whole scene.

Overall, LightOctree achieves a better compromise between realism, efficiency, and generalizability, which has potential for real-time lighting estimation in computationally constrained environments.

CONCLUSION

The development of indoor lighting estimation has progressed rapidly, from static global illumination models to the contemporary use of spatially-varying and learning-based methods. LightOctree is a recent example of this rapid advancement. It is a feasible and computationally efficient solution to getting the 3D lighting from only an RGB image. The development leverages a sparse voxel octree model and includes an amalgamation of a differentiable cone tracing renderer to represent indoor lighting in a scalable, distance-aware way that cuts down on memory and computation without sacrificing visual consistency.

Compared with previous methods such as DSGLight or StyleLight, which are accurate but rely on dense and slow architecture, Light Octree provides a more effective compromise between accuracy and performance. Its architecture can operate in real-time making it well suited for applications like augmented reality, cluttered scenes with insertion of virtual objects, or relighting of scenes on mobile or embedded systems where processing power is at a premium.

Beyond its current capabilities, the Light Octree framework opens promising avenues for further research and development. Its modular design lends itself well to interactive lighting editing, adaptive resolution rendering, and integration with generative models for user-guided illumination synthesis. Moreover, combining Light Octree with multimodal scene representations—incorporating geometry, materials, or semantics—could enable more holistic and intelligent scene understanding systems.

As immersive applications increasingly demand realism under constrained resources, the need for accurate, lightweight lighting models becomes ever more critical. Light Octree offers a timely and adaptable solution, bringing us closer to truly seamless integration between virtual and real environments.

REFERENCES

Wang, X., Xiao, S., & Liang, X. (2024). LightOctree: Lightweight 3D Spatially-Coherent Indoor Lighting Estimation. *CVPR 2024*.

Gardner, M.-A., Hold-Geoffroy, Y., Sunkavalli, K., Gagné, C., & Lalonde, J.-F. (2019). Deep Parametric Indoor Lighting Estimation. *ICCV 2019*.

Bai, J., Guo, J., Wan, C., Chen, Z., He, Z., Yang, S., Yu, P., Zhang, Y., & Guo, Y. (2022). Deep Graph Learning for Spatially-Varying Indoor Lighting Prediction. *arXiv preprint arXiv:2202.06300*.

Wang, G., Yang, Y., Loy, C. C., & Liu, Z. (2022). StyleLight: HDR Panorama Generation for Lighting Estimation and Editing. *arXiv preprint arXiv:2207.14811*.

Hilliard, J., Hilton, A., & Guillemaut, J.-Y. (2024). 360U-Former: HDR Illumination Estimation with Panoramic Adapted Vision Transformers. *arXiv preprint arXiv:2410.13566*.

Dastjerdi, A., et al. (2023). EverLight: Indoor-Outdoor Editable HDR Lighting Estimation. *ICCV 2023*.

Origami and Kirigami's Adaptive Potential in Architectural Shading Systems

An Up-to-Date Analysis of Responsive Geometries for Energy and Daylight Control

Ylleza Hoda[1]
[1]*TU Wien*
[1]*e12402551@student.tuwien.ac.at*

Origami and kirigami geometries offer unique opportunities for adaptive architectural shading systems because of their versatility, material efficiency, and spatial control. This report outlines three key studies on foldable and responsive facades to explore cutting-edge developments in this area. This report notes the material and technical hurdles yet to be conquered. It also promotes daylight modulation and energy savings when placed within the broader discourse on environmental simulation, smart materials, and kinetic design.

Keywords: *Origami, Kirigami, Adaptive Architecture, Responsive Facades, Dynamic Shading, Building Simulation, Smart Materials, Environmental Design.*

INTRODUCTION

Climate-responsive architecture, once viewed solely as passive solar design, has reached new heights with recent advances in computation, material technologies, and automation. Existing static components, such as louvers or overhangs, often are not enough to effectively mitigate fast-changing environmental conditions or spatial use changes. This gap has led to the development of systems with adaptive shading elements that can react in real-time to various parameters such as solar angles, temperature changes, and the preferences of building occupants.

As such, origami and kirigami as systems of geometry and design utilizing the folding and cutting of paper in Japan, have emerged as means to ascertain high degrees of spatial transformability, while avoiding excessive mechanical design, onto adaptable building skins; thus, providing surfaces that can bend, expand, contract, and transform in practice toward both functional and aesthetic ends.

The report begins with an exploration of origami- and kirigami-inspired shading systems in architecture today. The report also assesses three reviewed studies for their contributions, methods, and findings, and maps out the three studies in a wider interdisciplinary field consisting of computational geometry, smart materials, daylighting science, and energy modeling.

GEOMETRIC PRINCIPLES AND ARCHITECTURAL APPLICATIONS

The unique structural and kinematic spiriting of origami and kirigami geometries offer a set of parameters for consideration when creating adaptive architectural systems. Origami focuses on folding flat material into a three-dimensional shape. This process typically results in a compact structure that offers the potential for emergent self-supporting structures with embedded hinges. The inclusion of cuts or slits in the material are also important in kirigami, which facilitate not only new folded geometries but a more involved deformation that includes stretching, bending, and twisting; deformations that hinge on folding cannot reach (see Figure 1, adapted from Filipov et al., 2015).

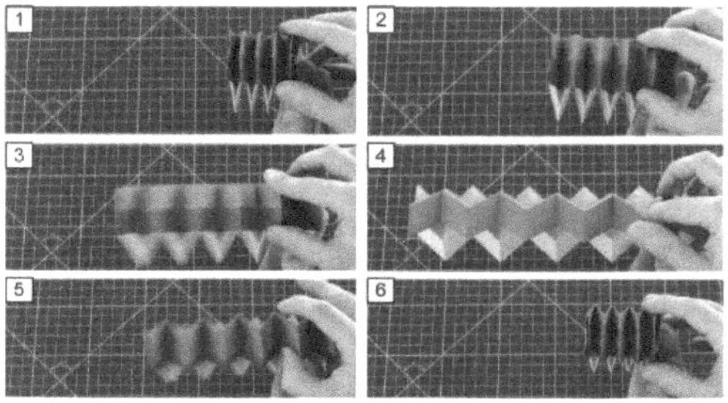

Figure 1 Comparison between Miura-ori origami folding and kirigami deformation patterns

The inherent mobility of both geometries is an obvious advantage in their architectural framework. Panels can be folded into a flat state and fixed together, which allows for the seamless transport of multiple panels that are then expanded on site, eliminating the challenges of fabrication with complex assembly points. There is also material efficiency from the geometries themselves, creating the potential to eliminate additional structural framing without losing structural integrity or support. Their ability to shape how the ratio of opening can be controlled, enables architects and designers to precisely modulate the solar exposure, glare, and heat gain within the building, which should improve energy performance without sacrificing occupant comfort or aesthetic value. The richness and dynamism of building surface is created through the repetition and transformation of folding patterns, which adds aesthetic value to the spaces they occupy.

These characteristics allow origami and kirigami systems to be used in a number of architectural purposes, including interior daylight screens, window shading panels, operable façades, pavilion envelopes, and double-skin cladding systems. Their incorporation into architectural design practice is greatly aided through computational means such as Grasshopper and Kangaroo, which provide a way to generate and test complex patterns, even a complex origami design. Environmental performance can be assessed using tools such as Radiance for daylight analysis, Energy Plus for thermal performance, and Abaqus for structural simulation, which results in a comprehensive workflow linking design intent and metrics.

SELECTED STUDIES: METHODS AND RESULTS

Interior Daylighting Screens – Nikookar & Omidfar Sawyer (2022)

Nikookar & Sawyer constructed a modular origami-inspired screen employing a rotational folding process, in which the animated properties of the system were to:

- Reduce direct sunlight
- Create an even illuminance throughout space
- Provide a distinct visual and sculptural component

They conducted Radiance simulations on the prototype classroom models, specifically measuring:

- Daylight Glare Probability (DGP)
- Spatial Daylight Autonomy (sDA)
- Annual Sunlight Exposure (ASE)

Outcomes:

- Some folding patterns were able to reduce glare by over 40%
- DGP was consistently below 0.35 during peak hours

- sDA improved by 18% from control conditions

Notably, their design was non-motorized and purely informed by geometry and material translucency. The study shows how non-kinetic origami can also operate within a passive adaptive system based on spatial disposition.

Kinetic Facade Panels – Pesenti et al. (2020)

The research was centered on a "dynamic origami"- based façade made of triangular panels using Miura-ori origami techniques. The authors developed parametric simulations and workflows that connected:

- Grasshopper for geometric generation
- Radiance for visual performance
- Energy Plus for heating/cooling loads

Actuation was realized by taking digital logic based on daily and seasonal solar angles, which would actuate the motors that adjust panel angles in real-time.

The results of this study were:

- Up to 25% cooling load savings in Milan and Rome, compared to baseline
- Improved daylight balance, without increasing glare
- Increased ability to block the high summer sun and admit the low winter sun, see Figure 2, adapted from Pesenti et al., 2020.

This paper illustrates the potential of kinetically responsive systems, where shape-shifting geometries are driven by algorithmic feedback loops. It represents origami, not just as an expression of form, but as a logic of climate control.

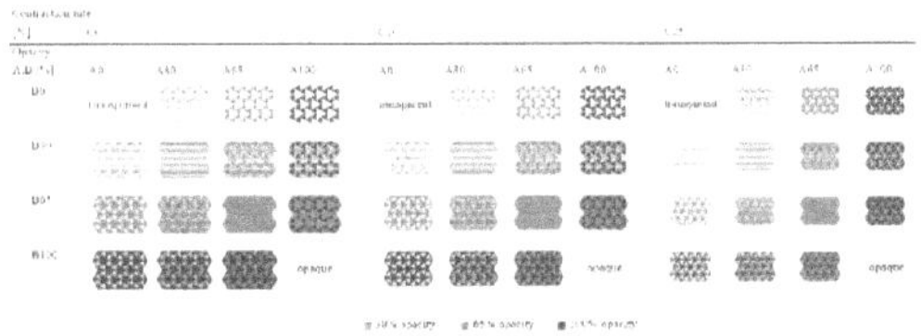

Figure 2 Simulation snapshots of adaptive origami façade panels in summer and winter configurations

Kirigami Façade Prototypes – Arauza et al. (2021)

This study explored passive thermal actuation. The authors utilized kirigami cuts in polymer sheets that expand and rotate when exposed to heat, altering the visual transparency or shading performance of the façade.

They did physical tests on their double-skin prototype using solar lamps and infrared cameras; their performance was modeled in energy analysis software.

Results:

- Surface temperature regulation & reduction in internal solar gains.
- Controlled deformation without mechanics.
- Visibility controlled through retention of partial aperture.

This system is unique in the form of its material intelligence: a smart material façade that "breathes" in the sun, distorting as a result of heat without mechanical systems. The authors validated this behavior through lab tests with solar lamps and thermal cameras which clearly showed the shift in aperture size and shape based on temperature at a higher rate of deformation, see Figure 3 adapted from Arauza et al., 2021. This makes the system especially promising for passive applications in high-solar-gain regions where maintenance requirements are low.

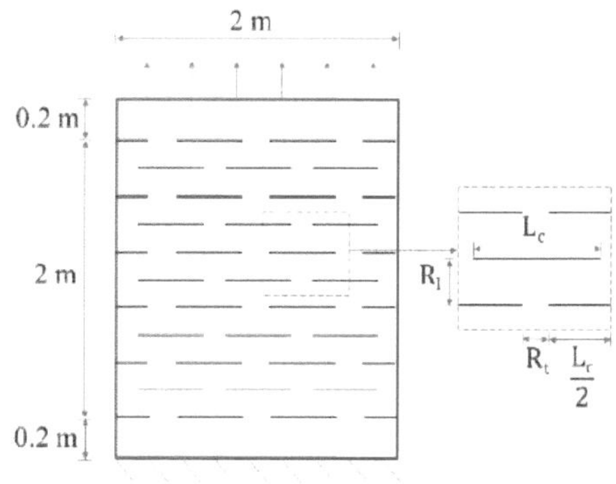

Figure 3 Photographs and thermal imaging of kirigami-inspired façade under heat exposure

CONCLUSION

The fusion of origami and kirigami techniques with digital technology and smart materials is leading to a new generation of adaptive architecture. These systems inherent in architecture, whether they are static screens, kinetic façades or passive responsive skins, pose exciting possibilities to reduce light and energy needs; a range of responses to climate change. Each individual approach has its own inherent compromises of complexity and performance; however, they collectively present additional approaches for adaptive or climate change-responsive design.

Going forward, the research agenda could address issues of real-time sensing, durability of materials, open-source design tools and hybrid approaches that combine folding logic and evolving technology. Ultimately, foldable geometries serve as a middle ground between form, function and environmental responsiveness, towards a more intelligent and adaptive built environment.

REFERENCES

Arauza, R., Filipov, E., Fascetti, A., Clifford, D.T. and Brigham, J.C. (2021). Building performance evaluation of a kirigami-inspired double-skin adaptive façade. Building and Environment, 202, p.108025.

Badarnah, L. (2015). Form follows environment: Biomimetic approaches to building envelope design for environmental adaptation. Buildings, 5(1), pp.287–303.

Filipov, E.T., Tachi, T. and Paulino, G.H. (2015). Origami tubes assembled into stiff, yet reconfigurable structures and metamaterials. PNAS, 112(40), pp.12321–12326.

Nikookar, N. and Omidfar Sawyer, A. (2022). Designing for daylight: An exploration of origami-inspired interior shading screens. LEUKOS, 18(2), pp.167–186.

Pesenti, M., Masera, G. and Fiorito, F. (2020). Exploration of adaptive origami shading concepts through integrated dynamic simulations. Journal of Building Performance, 11(1), pp.15–28.

Tachi, T. (2009). Generalization of rigid foldable quadrilateral mesh origami. Journal of the International Association for Shell and Spatial Structures, 50(3), pp.173–179.

Vasquez, J. and Wang, Z. (2021). Shape memory materials for responsive facades: A review. Energy and Buildings, 241, p.110961.

Zhai, Q., Sun, J. and Chen, X. (2022). Design and optimization of kirigami-inspired building skins with adaptive properties. Advanced Engineering Materials, 24(5), 2101199.

Machine Learning for Building Energy Optimization

A Review with a Focus on Occupant Behavior Modeling

Viet Loi Tran
Technische Universität Wien
e12347304@student.tuwien.ac.at

Buildings consume 40% of the world's energy, and occupant behavior significantly affects how efficiently they operate. With an emphasis on occupant behavior modeling, this paper summarizes current developments in machine learning (ML) for building energy efficiency. We examine data-driven methods such recurrent neural networks, reinforcement learning, and hybrid models that enhance occupancy forecasting, HVAC control, and retrofit decision-making through a comprehensive evaluation of peer-reviewed research published between 2023 and 2025. Case studies show that occupant-aware demand response and adaptive control techniques can result in energy savings of 20–30%. Future approaches for scalable and moral ML deployment are explored, along with issues like data quality, interpretability, and privacy. The results highlight ML's ability to balance occupant comfort and energy efficiency, hence promoting worldwide de-carbonization initiatives.

Keywords: *Machine Learning, Building Energy Optimization, Occupant Behavior Modeling, Reinforcement Learning, HVAC Control.*

INTRODUCTION

Heating, ventilation, and air conditioning (HVAC) systems are the second largest end-use, accounting for over 40% of global energy consumption in buildings (Li et al., 2025). Optimizing building operations is essential for cost savings and carbon reduction as urbanization and occupancy demands increase. Conventional physics-based simulation models are excellent for analyzing designs, but they may find it difficult to adjust in real time to changing conditions and occupant behavior. Machine learning (ML) techniques, on the other hand, make use of a wealth of sensor and operational data to predict energy use, deduce occupant trends, and inform adaptive control schemes. One of the biggest obstacles to building energy efficiency is still occupant behavior, which includes things like opening windows, regulating lights, and modifying thermostats. Actual building energy consumption may be up to 2.5 times greater than planned due to occupant activity uncertainty, which can cause significant disparities (Yu and Xu, 2024). Significant HVAC energy savings of more than 20% are possible with improved occupant awareness in control schemes (Fatehi et al., 2023). In the larger framework of building energy optimization, this paper concentrates on a few new data-driven algorithms for occupant behavior modeling. The objective is to summarize cutting-edge methods, look at real-world applications in case studies, and demonstrate how these discoveries fit into workflows for end-to-end optimization.

METHODOLOGY

We performed keyword searches in Google Scholar, IEEE Xplore, ScienceDirect, and MDPI for the years 2023–2025 in order to survey recent developments in data-driven occupant behavior modeling for building energy optimization. "Occupant behavior," "machine learning," and "building energy" were paired with particular algorithm names (such as "GRU" and "reinforcement learning") in the search phrases. We excluded works that were solely theoretical or physics-based and instead concentrated on peer-reviewed publications and credible conference papers that report on real-world implementations or case studies.

After conducting a preliminary search and screening, we found a core selection of research that demonstrated three applicability areas:

- Academic-building occupancy forecasting, exemplified by Fatehi et al. (2023), who compare GRU and LSTM networks for room-occupancy prediction using PIR motion-sensor logs;
- Office-building blind-control modelling, exemplified by Verma, Gopalakrishnan and Sonta (2023), who develop a blind-control occupant behavior model based on contextual and time-related factors using supervised machine learning approaches;
- Reinforcement-learning-driven occupant response models, as in Jang et al. (2023), who integrate pricing signals with a deep RL controller to learn and predict how occupants adjust HVAC setpoints in a workplace "game."

Information about model architectures, training methods, performance measures, data sources, data preprocessing, and feature engineering was taken out of each paper. These findings were then combined to show how behavior models are created, verified, and used in real-world energy-optimization processes. The review highlights methodological similarities (e.g., sequence modeling, classification techniques, reward design) and unique challenges (data privacy, real-time inference, occupant comfort trade-offs) in applying machine learning (ML) to occupant behavior modeling by focusing on these representative case studies.

MACHINE LEARNING IN BUILDING ENERGY OPTIMIZATION: A QUICK SURVEY

With data-driven insights for forecasting, control, and retrofit planning, machine learning (ML) has become a potent addition to physics-based simulation in building energy optimization. To set the stage for the more in-depth occupant behavior modeling that follows, we quickly go over three important application areas in this section: load forecasting, data-driven HVAC control, and retrofit decision support.

Load Forecasting

For scheduling, demand response, and system sizing, an accurate assessment of building energy demand is essential. Conventional statistical techniques can suffer from excessive dimensionality and nonlinearity and necessitate considerable expert tweaking. ML techniques, such as deep learning models (LSTM, CNN–LSTM hybrids) and tree-based ensembles (e.g., XGBoost, Random Forest), automatically extract features from historical consumption, weather, and calendar data to produce better forecasts (Li et al., 2025). By using data from several "source" buildings for data-scarce "target" sites, transfer learning further improves performance. Li et al. (2024), for example, demonstrate that a deep adversarial neural network may outperform untuned LSTM in prediction by 40% to 90%. Compared to parameter-optimized LSTM, gains of 20–80% allow for more dependable demand responsiveness.

Data-Driven HVAC Control

Building energy management requires HVAC systems. The first-principles thermal models that are commonly used in model predictive control (MPC) can be difficult to calibrate. In order to enable online optimization with little manual adjustment, data-driven MPC substitutes neural networks or ensemble regressors trained on previous operation data for these surrogates. Separately, HVAC control is formulated as a sequential choice issue using reinforcement learning (RL), which learns rules through interaction. In simulation, deep RL agents, like those that use Proximal Policy Optimization (PPO), integrated frameworks can lower temperature deviation by up to 14.32% and HVAC energy usage by 1.43–3.12% (Hou et al., 2024). By further integrating "planning guardrails," Jang et al. (2023) lower the exploration risk of RL and achieve 32% CO_2 savings in demand response scenarios via a synthetic model to assess possible actions prior to real-world trials. Deployment has been promoted due to transfer learning, which enables previously trained agents to adjust to new buildings with fewer training sessions.

Retrofit Decision Support

For the built environment to be deeply decarbonized, retrofitting existing structures is crucial, but it presents a combinatorial optimization difficulty. ML-based surrogate models, which are frequently artificial neural networks, are trained on the results of comprehensive simulation–optimization processes in order to forecast near-optimal retrofit options in terms of emissions and cost (Thrampoulidis, Hug & Orehounig, 2023). Using surrogate predictions and grouping buildings into archetypes, practitioners can quickly screen portfolios of hundreds or thousands of structures. Compared to brute-force simulation, this bottom-up method reduces the computational load by orders of magnitude without appreciably sacrificing accuracy. This makes it possible for facility managers and governments to rank actions across huge building stocks.

OCCUPANT BEHAVIOR MODELING FOR ENERGY EFFICIENCY

Significant fluctuation in building energy use can be introduced by occupant behaviors, such as plugging loads, opening windows, and altering thermostats. Using machine learning to model these patterns allows for more responsive control tactics in addition to increasing forecast accuracy. In this section, we outline the standard preprocessing procedures and data sources, provide an overview of the primary algorithmic techniques, and highlight three representative case studies.

Data Sources and Preprocessing

Common inputs for behavior models include:

- Motion sensors (PIR): Provide binary or count-based signals indicating presence in rooms or zones.
- Environmental sensors: Temperature, CO_2, and light-level readings correlate with occupant comfort preferences and actions.
- Digital traces: Wi-Fi or badge-swipe logs infer arrival/departure schedules.
- Imaging or contact sensors: Cameras or reed switches detect window and door states.

In preprocessing, multi-sensor streams are usually synchronized at a consistent sampling rate (e.g., hourly or 5-minute according to Li et al., 2024), missing data is imputed (e.g., through forward-fill or interpolation), and features like time of day, indoor-outdoor temperature differentials, and recent occupancy history are derived. Scaling or standardization guarantees input compatibility with neural network topologies.

Algorithmic Approaches

According to the reviewed literature, occupant behavior modeling spans three main ML paradigms:

- Time-Series Forecasting: Based on past sequences, recurrent networks (LSTM, GRU) and attention-based models are utilized to forecast future occupancy or energy loads. GRUs are appropriate for workloads requiring quick, scalable forecasting since they frequently contain fewer parameters, train more quickly, and have equivalent or slightly higher accuracy than LSTM.
- Classification & Clustering: For classification tasks like fault detection and forecasting discrete occupant actions like blind states, supervised classifiers like Random Forest and SVM are used. K-Means and DBSCAN are examples of unsupervised clustering techniques that find patterns in data to aid in problem diagnosis and scenario analysis of energy use.
- Probabilistic & Hybrid Models: Stochastic transitions between states are captured by Markov chains, which are used to represent occupant behavior. To handle complicated scenarios and solve data limit, hybrid approaches incorporate various learning approaches, such as supervised and unsupervised techniques or deep learning models.

Acquiring the trust of operators and occupants requires explaining model judgments using SHAP or LIME, particularly when behavior predictions activate automated controls.

Case Studies

University Building PIR Forecast: Fatehi et al. used motion-sensor data from university buildings' classrooms for ten weeks to train both LSTM and GRU networks. With 13.57% less parameters, the GRU obtained a somewhat lower Mean Average Error (MAE), indicating that lighter architectures may be better suited for quick occupancy predictions. Demand response and energy efficiency applications in smart buildings are supported by their model.

Blind Control Occupant Behavior Model (BC-OBM): Verma, Gopalakrishnan, and Sonta created a data-driven algorithm that uses just contextual and time-related factors (such as orientation, WWR, blind type, and ventilation mode) to predict office blind placements. They gathered 935 observations from 42 offices in India over the course of six months. Random Forest (RF) and mixed-effect binary logistic regression (BMLR) were the two models that were examined. The BMLR model outperformed the RF (73.4%) in terms of accuracy (84.7%) and variance explanation ($R2 = 0.422$). The main predictors were orientation, WWR, and ventilation mode.

RL-Driven Occupant Response Models: Jang et al. combined simulated occupant "pricing games" with a deep reinforcement learning agent. The RL policy adjusted its energy consumption behavior to learn how workers responded to dynamic price signals. Their risk-aware strategy demonstrated its potential for demand response by influencing human behavior and reducing deployment risk, as seen by the 30% energy cost reductions relative to time-of-use rates when implemented in a simulated medium-sized office complex.

Integration into Energy Optimization Workflows

Behavior models inform downstream optimization in two primary ways:

- Forecast-Driven Control: Model predictive control loops are fed by predicted occupancy schedules, which enable HVAC setpoints to proactively adjust to anticipated presence.
- Occupant-Aware Demand Response: Tailored incentive signals are made possible by real-

time action predictions, which guarantee that demand-response events coincide with expected occupant compliance and have the least possible negative influence on comfort.

This machine learning models help buildings function more effectively while preserving user satisfaction by bridging the gap between behavior prediction and control.

CHALLENGES AND FUTURE DIRECTIONS

Data Quality and Transferability

Sensor data is crucial for occupant behavior and building energy models, but buildings can experience data scarcity situations due to inadequate or missing data. Incomplete, noisy, or missing data can cause the model to perform worse. Transfer learning (TL) transfers knowledge from data-rich "source" buildings to data-poor "target" buildings, solving the data shortage problem. Choosing appropriate source buildings is still a major challenge in optimizing TL benefits. Future research investigates automated similarity measures like Dynamic Time Warping (DTW) for time series data and creating synthetic data to strengthen model robustness in the face of data constraints.

Interpretability and Trust

"Black box" models like artificial neural networks and reinforcement learning agents are energy-efficient and accurate, but their lack of transparency hinders widespread use. Interpretability is crucial for decision-makers and building operators to understand model predictions. Explainable AI (XAI) techniques like SHAP and LIME reveal how features contribute to model predictions. Efforts to enhance acceptance and reliability should focus on developing resilient, scalable, and flexible machine learning frameworks that incorporate XAI techniques.

Multi-Objective Trade-Offs

A number of goals must be balanced to optimize building energy systems, including comfort, cost, energy, and CO_2 emissions/carbon intensity. Surrogate models help anticipate the best solutions when multi-objective optimization is used for building retrofit and planning to balance cost and carbon emissions. Reward functions in RL-based control balance occupant comfort and energy economy by imposing penalties for temperature variations. To better balance these goals and adjust to occupant needs, future research will focus on creating more flexible machine learning frameworks and improving reward functions.

Privacy and Ethical Considerations

Concerns about data ownership and surveillance are brought up by fine-grained behavior modeling. Although they require strong protocols to prevent model inversion attacks, techniques like federated learning—training shared models across numerous buildings without centralized data pooling—can reduce privacy threats. Future research should provide ethical guidelines for the use of occupant data and standardize anonymization pipelines to ensure informed consent and clear value exchange (e.g., occupants receive comfort enhancements in exchange for data).

CONCLUSIONS

Machine learning has become a game-changing tool for solving important building energy efficiency problems, especially when it comes to occupant behavior modeling. Li et al. (2024) have shown that recent developments in forecasting models, including neural networks and ensemble methods, have attained noteworthy accuracy and generalizability using techniques like transfer learning. Studies like Jang et al. (2023) have reported energy savings of up to 30% while preserving occupant comfort, demonstrating the great potential for real-time HVAC optimization with data-driven control methodologies like model predictive control (MPC) and reinforcement learning (RL). In order to facilitate proactive energy management, occupancy behavior models that make use of time-series forecasting and hybrid algorithms improve the correspondence between human behavior and system reactions.

Multi-objective trade-offs, model interpretability, and data scarcity continue to be significant obstacles. To guarantee scalability and stakeholder trust, future research should give top priority to explainable AI methods, strong transfer learning frameworks, and ethical data practices. By combining these developments, ML-driven methods can close the gap between expected and actual energy performance, hastening the built environment's de-carbonization.

REFERENCES

Fatehi, N. et al. (2023) Machine Learning based Occupant Behavior Prediction in Smart Building to Improve Energy Efficiency, p. 5. Available at: https://doi.org/10.1109/ISGT51731.2023.10066411 (Accessed: 13 May 2025).

Henkel, P. et al. (2024) 'Interpretable data-driven model predictive control of building energy systems using SHAP', in Proceedings of the 6th Annual Learning for Dynamics & Control Conference. 6th Annual Learning for Dynamics & Control Conference, PMLR, pp. 222–234. Available at: https://proceedings.mlr.press/v242/henkel24a.html (Accessed: 13 May 2025).

Hou, F. et al. (2024) 'Multi-source transfer learning method for enhancing the deployment of deep reinforcement learning in multi-zone building HVAC control', Energy and Buildings, 322, p. 114696. Available at: https://doi.org/10.1016/j.enbuild.2024.114696 (Accessed: 13 May 2025).

Jang, D. et al. (2023) 'Deep reinforcement learning with planning guardrails for building energy demand response', Energy and AI, 11, p. 100204. Available at: https://doi.org/10.1016/j.egyai.2022.100204 (Accessed: 13 May 2025).

Li, D. et al. (2025) 'Machine Learning Applications in Building Energy Systems: Review and Prospects', Buildings, 15(4), p. 648. Available at: https://doi.org/10.3390/buildings15040648 (Accessed: 13 May 2025).

Li, G. et al. (2024) 'Comprehensive transferability assessment of short-term cross-building-energy prediction using deep adversarial network transfer learning', Energy, 299, p. 131395. Available at: https://doi.org/10.1016/j.energy.2024.131395 (Accessed: 13 May 2025).

Thrampoulidis, E., Hug, G. and Orehounig, K. (2023) 'Approximating optimal building retrofit solutions for large-scale retrofit analysis', Applied Energy, 333, p. 120566. Available at: https://doi.org/10.1016/j.apenergy.2022.120566 (Accessed: 13 May 2025).

Verma, T., Gopalakrishnan, P. and Sonta, A. (2025) 'Development of a blind control occupant behaviour model (BC-OBM) based on contextual and time-related factors through supervised machine learning approaches', Building and Environment, 277, p. 112932. Available at: https://doi.org/10.1016/j.buildenv.2025.112932 (Accessed: 13 May 2025).

Yu, H. and Xu, X. (2024) 'Reinforcement learning for occupant behavior modeling in public buildings: Why, what and how?', Journal of Building Engineering, 96, p. 110491. Available at: https://doi.org/10.1016/j.jobe.2024.110491 (Accessed: 13 May 2025).

Path Finding Algorithms

Emir Softic

TU Wien

e12425770@student.tuwien.ac.at

Path planning is a fundamental aspect of robotics and automation, playing a critical role in enabling autonomous systems—such as mobile robots, UAVs, and self-driving cars—to navigate efficiently and safely from a source to a destination. As the demand for high-speed, intelligent systems grows, various algorithms have been developed to address the challenges of determining optimal, collision-free, and cost-effective paths. This paper categorizes path planning algorithms based on their application domains (e.g., pedestrian dynamics, emergency planning, urban design), algorithmic principles (exact, sampling-based, and machine learning-based), and key properties such as environment type (static or dynamic), algorithm scope (global or local), and completeness. Widely used algorithms—including Dijkstra's, A, Floyd-Warshall, RRT, and reinforcement learning methods—are analyzed in terms of their assumptions, strengths, and limitations across different environments. The overview emphasizes that no single algorithm universally outperforms others; rather, algorithm selection must align with the specific use case and environmental constraints. This comparative study serves as a guide for choosing suitable path planning techniques across various domains of robotics and automation.*

Keywords: eCAADe Algorithms, Path finding

INTRODUCTION

According to Gasparetto et al (2015), path planning is crucial issues in the field of Robotics and, more generally, in the field of Automation. Indeed, the trend for robots and automatic machines is to operate at increasingly high speed, in order to achieve shorter production times. . With the development of autonomous systems like mobile robots, unmanned aerial vehicles and autonomous cars in order to identify safe, efficient, collision-free and least-cost travel paths from an origin to a destination. There are a lot of different algorithms which find the shortest paths, but the way of how they do this and assumptions behind them differed, giving us as the result that no algorithm is best in any case, but that we need to provide different algorithms for different types of problems.

Path planning categorization

Path algorithms can be categorized in multiple ways, depending on the context, especially when applied to architecture and planning. Even tho Liao et al (2025) gives detailed categorization that should be studied, I will briefly sum up this. In terms of planning, we can categorize algortihtms by their application in different fields. There we would have:

1. Pedestrian Dynamics which can be macroscopis(flow optimisation like metro station layouts, using Dijkstra), and microscopic(agent-based modeling, like social forte models or using A* for evacuation)
2. Structural/Urgan Design giving service paths and generative design.
3. Emergency Planning is concerned with evacuation routes(usually doing dynamic replanning with RRT* algorithm).

Path planning algorithms can also be categorized by algorithm type. With this type of categorization we would have:

1. Exact algorithms that guarantee oprimality(like Dijkstra's algorithm, A* or floyd-warshall) 5. Approximate/Sampling-based which are really scalable and useful for xomplex spaces(such as Genetc algorithms and RTT) 6. Machine Learning algorithms.
2. Approximate/Sampling-based which are really scalable and useful for xomplex spaces(such as Genetc algorithms and RTT)
3. Machine Learning algorithms.

We can further devide path algorithms by heuristic focus, but this is more relevant for papers in Computer Science, since they are primarily discussing different algorithms, not discussing different algorithms based on their computation power.

Properties of Path Planning

Mobile robot path planning, as well as any other type of path planning can be characterized by three key properties, being environment type, algorithm scope and completeness. Environment type can be static, meaning that the environment contains no moving obstacles, and dynamic, where environment includes moving or changing obstacles like other pedestrians, robots, etc. Algorithm scope can be global, where a priori knowledge of the environment exists in terms of maps and grids in order to compute an optimal path upfront, as well as local planning algorithms, where we must rely on real-time sensor input for obstacle avoidance and trajectory generation. Lastly, completeness of our solution, which can be complete, meaning that the algorithm guarantees a solution if it exists, and heuristic, where algorithms sacrifices these guarantees for computational efficiency. If mobile robot navigation problem is our concerne, that this can be devided into three subtasks, namely mapping and modeling the environment, path planning and path traversal with collision acoidance, according to Behnke (2004).

List of Path Planning algorithms by use case

Now we try to give a list of algorithms based on their application and properties that were already described. Static Environment has three main algorithms:

Dijkstra's algorithm solves the porblem of finding the shortest path from the source to a destination. With this algorithms, we can find the shortest paths from a given source to all points in environment in the same time, which is the reason why this algorithm is sometimes called the single-source shortest paths problem according to Javaid (2013). This algorithm guarantees optimal paths in static grids or maps, making it useful for metro station layouts, utility routing, etc. Since it is determening the shortest paths to all destinations, it gets computationaly expensive for large spaces.
A-star(A*) algorithm is a heuristic search algorithm made as an improvement of Dijkstra algorithm. The heuristic search is implemented through an intelligent evaluation function f(x) at each node x: $f(x) = g(x) + h(x)$, where g(x) represents the exact cost function from the start node to the current node x, and h(x) denotes the

heuristic estimate of the remaining cost from x to the goal node, which was proposed by Tang and Ma (2021). This heuristic estimate can take several formes, with common implementations including Euclidean and Manhattan distance. This algorithm is optimal pathfinding with usage of heauristics, but requires precomputed maps, struggling with dynamic obstacles.

Floyd-Warshall algorithm, described by Bukhori and Dengen (2018), is a dynamic programming approach type that solves all-pairs shortest path problems by systematically evaluating every possible route through intermediate nodes. Is builds solutions incrementally using previous subproblem solutions and maintains two nxn matrices tracking minimal distance between nodes and intermediate nodes in optimal paths. With this algorithm, we can handle negative weights(but not negative cycles) returinging both shortest paths and actual paths. This algorithm is really effective, but has high memory usage.

Machine Learning based path planning is based on ML/AI fields with algorithms:

Reinforcement Learning algorithms which use agents, which actions are rewarded, with a goal of a agent to have a maximum cumulative sum of rewards one the task is finished. This is really useful in adaptive navigation in unpredictable environments like crowd evacuation but requires extensive training data.

Graph Neural Networks(GNNs) are described in Distils post as optimizable transformation on all attributes of the graph which usually used message passing neural network as a framework. Primarily they are used in learning spatial relationships, but are computationally intensive.

CASE STUDY: AUTONOMOUS WAREHOUSE ROBOTS

In modern automated warehouses, fleets of mobile robots navigate through obstacle-rich aisles to transport goods from storage to packing stations. A common architecture combines a global planner—often based on A* due to its optimality and ease of implementation on grid maps—with a local replanner such as the Dynamic Window Approach (DWA) for real-time obstacle avoidance around humans and other robots Nature. Based on Xiang at al (2022) and Fan et al (2024), researchers have shown that integrating A* for coarse route computation with a sampling-based method like RRT* can reduce overall travel time by up to 15% while maintaining collision-free guarantees. For task coordination among multiple robots, Conflict Based Search (CBS) layered on top of individual A* planners ensures deadlock-free multi-agent paths, trading off a modest increase in planning time for global completeness. Another important study is urban drone delivery which operates in three-dimensional space, with global pathing usually being done by A*, and for dynamic replanning the Informed RTT* is responsible, which was described in paper by He et al (2025).

REFERENCES

Gasparetto, A., Boscariol, P., Lanzutti, A. and Vidoni, R., 2015. Path planning and trajectory planning algorithms: A general overview. Motion and operation planning of robotic systems: Background and practical approaches, pp.3-27.

Liao, Y., Xu, X., Bai, R., Yang, Y., Cao, M., Yuan, S. and Xie, L., 2025. Following Is All You Need:

Robot Crowd Navigation Using People As Planners. arXiv preprint arXiv:2504.10828.
Behnke, S., 2004. Local multiresolution path planning. In RoboCup 2003: Robot Soccer World Cup VII 7 (pp. 332-343). Springer Berlin Heidelberg.
Javaid, A., 2013. Understanding Dijkstra's algorithm. Available at SSRN 2340905.
Tang, Z. and Ma, H., 2021, July. An overview of path planning algorithms. In IOP conference series: earth and environmental science (Vol. 804, No. 2, p. 022024). IOP Publishing.
Bukhori, D. and Dengen, N., 2018, April. Floyd-warshall algorithm to determine the shortest path based on android. In IOP Conference Series: Earth and Environmental Science (Vol. 144, No. 1, p. 012019). IOP Publishing.
Lei, S., Li, T., Gao, X., Xue, P. and Song, G., 2025. Research on improved RRT path planning algorithm based on multi-strategy fusion. Scientific Reports, 15(1), p.13312.
Distil - https://distill.pub/2021/gnn-intro/
Xiang, D., Lin, H., Ouyang, J. and Huang, D., 2022. Combined improved A* and greedy algorithm for path planning of multi-objective mobile robot. Scientific reports, 12(1), p.13273.
Fan, H., Huang, J., Huang, X., Zhu, H. and Su, H., 2024. BI-RRT*: An improved path planning algorithm for secure and trustworthy mobile robots systems. Heliyon, 10(5).
He, X., Li, L., Mo, Y., Sun, Z. and Qin, S.J., 2025. Air Corridor Planning for Urban Drone Delivery: Complexity Analysis and Comparison via Multi-Commodity Network Flow and Graph Search. Transportation Research Part E: Logistics and Transportation Review, 193, p.103859.

"A Review of Algorithmic Planning Methods for Building Envelope Retrofit Optimization"

Arman Kafashi[1]
[1]*Technical University of Vienna*
[1]*e12346668@student.tuwien.ac.at*

This paper reviews recent advancements in algorithmic planning methods used for building envelope retrofit optimization. Given that buildings contribute significantly to global energy consumption and greenhouse gas emissions, improving the energy efficiency of the existing building stock is crucial. Among the various retrofit strategies, building envelope interventions are particularly impactful. This review focuses on heuristic and simulation-based planning approaches, deliberately excluding artificial intelligence and machine learning methods. Recent literature (2023–2025) is systematically analyzed to highlight the primary tools, methodologies, evaluation metrics, and practical limitations in the application of these methods. The paper concludes by identifying key research gaps and proposing future research directions.

Keywords:. *Envelope retrofitting, Simulation-based design, Building performance*

INTRODUCTION

Buildings are responsible for approximately 36% of global final energy consumption and nearly 40% of energy-related carbon dioxide emissions (IEA, 2023). Consequently, retrofitting the existing building stock has emerged as a vital strategy to enhance energy efficiency and support climate goals. Among the retrofit options available, improvements to building envelopes—including walls, roofs, windows, and insulation—offer a cost-effective and technically feasible solution.

However, planning effective retrofit interventions presents a range of challenges, including limited data availability, complex trade-offs between cost and performance, and multiple stakeholder objectives. To address these challenges, a growing body of research has focused on algorithmic planning techniques that support systematic evaluation of retrofit strategies. Notably, these techniques can guide practitioners in balancing conflicting priorities, such as energy savings, thermal comfort, cost-effectiveness, and occupant well-being.

This paper reviews recent contributions (2023–2025) to the field of envelope retrofit planning using non-AI algorithmic methods. The review seeks to explore the following research questions:

1. What heuristic and simulation-based methods have been applied in building envelope retrofit planning?
2. Which tools and frameworks are used to implement these methods?
3. What criteria are used to evaluate the performance of these planning approaches?
4. What are the main limitations, and how can future research address them?

Building Envelope Retrofits: Importance and Challenges

The building envelope serves as the primary barrier between indoor and outdoor environments, significantly influencing thermal performance, comfort, and energy demand. Retrofitting envelope components has demonstrated substantial potential in reducing space heating and cooling loads, particularly in older buildings with outdated materials and construction methods (Li et al., 2024).

Typical envelope retrofit measures include:

- Increasing insulation levels
- Installing high-performance glazing systems
- Sealing air infiltration points
- Enhancing thermal mass and external shading

Despite their benefits, envelope retrofits face numerous barriers. These include high upfront costs, uncertain payback periods, and variability in building-specific conditions. In addition, the trade-offs between competing goals (e.g., energy savings vs. occupant comfort) complicate decision-making. Algorithmic planning methods provide structured, data-driven tools to evaluate and optimize these retrofit strategies.

Algorithmic Planning Methods for Retrofit Optimization

This section outlines key non-AI algorithmic methods currently employed in building envelope retrofit planning.

Heuristic Optimization Techniques

Heuristic optimization methods such as Genetic Algorithms (GAs), Particle Swarm Optimization (PSO), and rule-based decision trees are widely used in retrofit planning. These methods support multi-objective optimization by identifying cost-effective combinations of retrofit measures while satisfying thermal comfort, resilience, and emissions reduction goals. For instance, Li et al. (2024) introduced a GA-based framework that incorporates a Thermal Resilience Index to evaluate both efficiency and robustness under power outage scenarios.

Simulation-Based Planning

Simulation-based methods integrate performance modeling tools (e.g., EnergyPlus, TRNSYS) into the planning process. These tools simulate energy performance under varying climatic conditions, building usage patterns, and system configurations. Cai et al. (2025) developed a simulation-driven framework that evaluates thousands of retrofit combinations tailored to building-specific attributes, enabling informed envelope upgrade decisions.

Rule-Based and Bi-Level Planning

Rule-based systems encode expert knowledge and regulatory standards into structured decision frameworks. Ran et al. (2023) proposed a bi-level optimization model for integrated energy systems, where high-level capacity planning is separated from operational strategy optimization. This logic can be applied to envelope planning by combining strategic material and configuration selection with operational assessments.

Tools, Frameworks, and Case Studies

A range of tools and planning frameworks have been developed to support envelope retrofit decisions:

- **EnergyPlus**: A simulation engine widely used for modeling thermal behavior under dynamic conditions.

- **BEopt**: Developed by NREL, BEopt evaluates residential retrofit packages using life-cycle cost minimization.
- **DesignBuilder**: Provides a user-friendly interface for parametric simulation and performance assessment.
- **RETScreen**: Enables preliminary feasibility analysis and environmental impact assessments.
- **Custom Frameworks**: Cai et al. (2025) introduced a retrofit planning engine combining energy simulations with heuristic optimization tailored to specific building types and climates.

Case Studies:

- In Shanghai, GA-based optimization applied by Li et al. (2024) resulted in 25–65% reductions in energy use and improved comfort.
- Cai et al. (2025) demonstrated significant energy savings through tailored envelope retrofits in Canadian residential buildings.
- Zhou et al. (2024) applied a shoebox modeling approach to simulate district heating loads, integrating occupancy dynamics for better retrofit prioritization.

These case studies demonstrate the versatility of algorithmic frameworks across different climates, building types, and project scales.

Evaluation Criteria and Limitations

Evaluation of algorithmic retrofit planning methods typically involves the following indicators:

- **Annual energy savings**: Expressed in kWh/m^2 or percentage reduction from baseline.
- **Economic metrics**: Including payback periods, life-cycle costs, and cost-benefit ratios.
- **Thermal comfort**: Measured via operative temperature ranges or comfort indices such as PMV/PPD.
- **Environmental performance**: Reflected through CO2 emissions or primary energy use.
- **Resilience**: Evaluated based on system behavior during power outages or thermal disruption.

Limitations and Challenges:

- **High computational demand**: Exhaustive simulations increase modeling time and resource use.
- **Tool fragmentation**: Existing tools often address isolated performance aspects (e.g., cost only or energy only).
- **Uncertainty and variability**: Occupant behavior, weather fluctuations, and data quality introduce unpredictability.
- **Usability**: Complex tools may not be accessible to non-experts or early-stage designers.
- **Integration barriers**: Limited interoperability with BIM platforms or building audit databases hinders adoption.

Future developments should prioritize user-friendly interfaces, stakeholder-inclusive workflows, and the integration of multiple evaluation criteria into a single tool.

Conclusions and Future Directions

Algorithmic planning approaches for building envelope retrofits offer structured, data-informed pathways to improve energy efficiency, thermal performance, and climate resilience. This review has shown that heuristic and simulation-based methods, though not reliant on artificial intelligence, are highly capable of supporting retrofit decision-making when tailored to building-specific conditions.

Nevertheless, several limitations persist, including the fragmented nature of available tools, the complexity of managing multi-criteria optimization, and limited practical adoption. Further research is needed to develop integrated frameworks that can synthesize cost, energy, comfort, and emissions criteria into actionable retrofit strategies.

Priority areas for future research include:

- Combining rule-based and simulation-driven planning methodologies
- Developing modular, easy-to-use interfaces for non-expert users
- Incorporating district-level planning and adaptation to future climate conditions

Ultimately, refining these approaches will support the broader goal of transforming the existing building stock into a more sustainable and resilient component of the built environment.

Acknowledgment

This paper was language-edited and paraphrased with the assistance of ChatGPT, an AI language model by OpenAI.

References

Cai, J., Rayegan, R., Aryal, A. (2025). An Automated Framework for Tailored Building Envelope Retrofit Planning. *Journal of Building Engineering*. (In press).

IEA (2023). Global Status Report for Buildings and Construction. *International Energy Agency*. https://www.iea.org/reports

Li, Y., Zhang, Y., & Wu, Q. (2024). A New Method of Building Envelope Thermal Design Optimisation Combined with Room Thermal Resilience Index. *Journal of Building Physics*, 47(3), 267–285.

Ran, X., Liu, J., & Fang, X. (2023). A Bi-Level Optimization Method for a Regional Integrated Energy System Under Climate Change Conditions. *Proceedings of Building Simulation 2023*, 172–181.

Zhou, W., He, J., & Liu, Z. (2024). A New Method for Dynamic Load Simulation of Urban Building Complexes' Heating Considering Building-Level Occupancy Based on Shoebox Algorithm. *Building and Environment*, 244, 110948.

Efficient Emergency Exits

A problem of Emergency Exits Layout

Ângelo Miguel Pereira do Carmo
Technische Universität Wien
e11722580@student.tuwien.ac.at

The Emergency Exit Layout Problem (EEPL) focusses on optimizing emergency exits locations and improving safety in large public venues. An agent-based simulation is used to model crowd behavior during emergencies, helping to understand the effects of exit placement, crowd distribution, and incident scenarios on evacuation dynamics. A two-stage stochastic mixed integer linear programming model (MILP) integrates pre-incident design decisions, and post-incident crowd behavior, and finally, an agent-based simulation evaluated the effectiveness of the layout. Two study cases are analyzed, a large outdoor concert and a small indoor arena, demonstrating that exist width is critical in open spaces, and detailed risk assessment and scenario modeling are essential on complex environments.

Keywords: *Emergency Exits Layout Problem, EEPL, MILP, Outdoor Venues, Indoor Venues, Agent-Based Simulation, Optimization.*

INTRODUCTION

Planning for safe evacuations is crucial, especially in large public spaces where safety is a priority. One of the biggest considerations for ensuring safety is with the arrangement of emergency exits in terms of quantity and location, as this has a direct impact on the distances and time needed for evacuating. The report analyzed considers a challenge called the Emergency Exit Layout Problem (EEPL), which involves using mathematical modelling and computer simulation to study the evacuation of crowds in an emergency. It assesses two types of real-world situations, an outdoor large-scale concert, and an indoor smaller concert. Both situations were examined, and practical conclusions were drawn about the effects of better planning. The case studies show how the placement of the crowds in the environment and the location of incidents influence the overall safety and effectiveness of the exits and the value of having in-depth plans in place, prior to an emergency happening.

DEFINE THE EMERGENCY EXIT LAYOUT PROBLEM (EEPL)

The Emergency Exits Layout Problem (EEPL) helps with planning and optimizing safety during emergency evacuations of large spaces and events, such as concerts, sports venues, or gatherings. The EEPL helps determining the optimal number, location and width of emergency exits, along the boundary of a space, or area used for events, and to minimize the total evacuation time required for a crowd to reach safety.

The EEPL considers several elements:

- Space geometry: the physical layout, or geometry of the venue, where emergency exits can be, and cannot be placed.
- People distribution: the number and the location of people within the space at the time of an incident.
- Incident scenarios: occurrences that can trigger an evacuation of the premises, such as a fire, and explosion, or another time of alarm, each with different impact zones, and line-of-sight obstructions.
- Exit capacity: exit flow rate is modeled as a linear function of the width. Exits are built with a predefined number of width units.

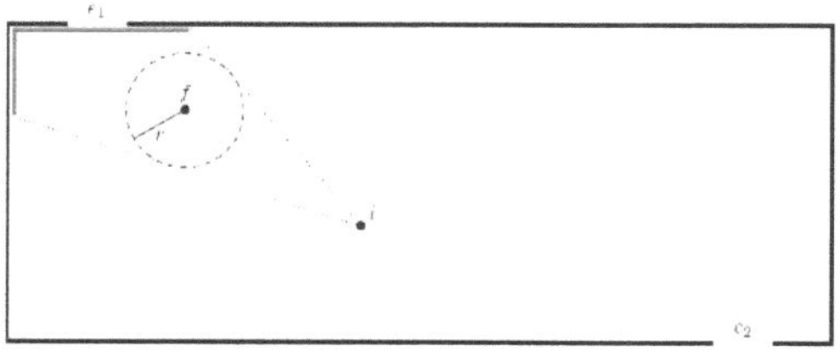

Figure 1
Fire f, obstructing an emergency exit.

This is modeled as a two-stage stochastic mixed-integer linear program (MILP). In the first stage, design decisions made before an incident occur are modeled, specifically, choosing exits and specifying widths under a limited total width constraint. As for the second stage, real time decisions made by people during an evacuation are modeled, such as, which exits people should use and how the crowds flow through the space. In this second stage, it is included a set of possible scenarios, representing combinations of people distribution and incidents.
The EEPL unifies infrastructure layout and behavioral assumptions, and this allows planners and safety officials to analyze different situations and prepare for a range of emergencies, optimizing safety in practical terms.

MODEL THE EEPL USING OPTIMIZATION

To solve the EEPL, the problem is modeled and optimized as a MILP. The framework is designed to integrate pre-event design decisions and the post-incident operational responses, and incorporating uncertainties related to people distribution in the area of events, and incident scenarios.

In the first stage, the model makes design decisions, specifically:

- Where to locate emergency exits along the boundary.
- How wide should each emergency exit be, using a modular width system with fixed total width constraints, across all exits.

The width of the emergency exits is defined by modules: "The total width is divided into equally sized modules, and an integer number of modules can be placed at an emergency exit. The number of modules allocated to different exits can vary." (Section 2 – Problem description and formulation), and the total amount of modules available is limited. The width of each emergency exit has a direct impact on the flow rate of each exit, therefore limiting the number of people that can exit per module per time period.

In the second stage, it's simulated how people respond to a given incident. By computing:

- Flow of people from different zones withing the space, to each exit.
- Queue dynamics at exits, calculating the number of people waiting and exiting overtime.
- Evacuation time is the time required for a specified number of people to leave the premises.

Each modeled scenario is a combination of incident location, and people distribution at the time of the incident. The scenarios are assigned probabilities, and the goal of the optimization process is to minimize the evacuation time, weighted across all scenarios.

The model includes several constraints to ensure feasibility and realism, for example:

- A fixed number of exits.
- The total width cannot exceed a predefined limit.
- No people can be assigned to non-existent exits.
- Queues and flows must respect the capacity constraints defined by exit width.
- People must leave the venue through exits that are reachable (not obstructed by fire, or other obstacles).

The model can also restrict people behavior by assuming that each person uses the nearest visible exit, and two solutions are tested: Time-Centered (TC), which prioritizes minimizing evacuation time, and Distance-Centered (DC), focused on reducing the distance traveled.

EVALUATION WITH AGENT BASE SIMULATION

To evaluate the emergency exits location proposed by the optimization model, a detailed agent-based simulation model is used. The goal is to verify whether the optimized locations are effective under a variety of emergency scenarios. This simulation serves to replicate actual human behavior during evacuations and assess total evacuation time, distance, exit usage, or other parameters.

Simulation Objectives and Structure

The simulation is designed to replicate the dynamics of an evacuation of a single floor arena with a specified number of static obstacles, and various emergency scenarios. The space is modeled as a 2D continuous area, allowing people to move in any direction.

- Each person is represented as an agent with attributes, including:
- Personality type: with Leader, Follower and Panic personalities influencing exit choice.
- Heath status: Unharmed, Minor Injuries, Urgent, Acute, or Deceased.
- Movement characteristics: Speed, Direction and Position.

Health status affects behavior, for example, injured or deceased agents become static obstacles for other agents, making movement patterns more complex.

Three-Level Decision Process

The agent's behavior is controlled by a hierarchical decision-making process with three levels.

- Strategic level: choosing an exit based on utility. Factors like distance, crowd density, obstacles, and decisions of nearby agents are considered.
- Tactical level: chooses the shortest path to the chosen exit and adapts for blocked routes and crowd density.
- Operational level: the next step is calculated using social-force modeling, which simulates physical and psychological influences from the environment and from the other agents.

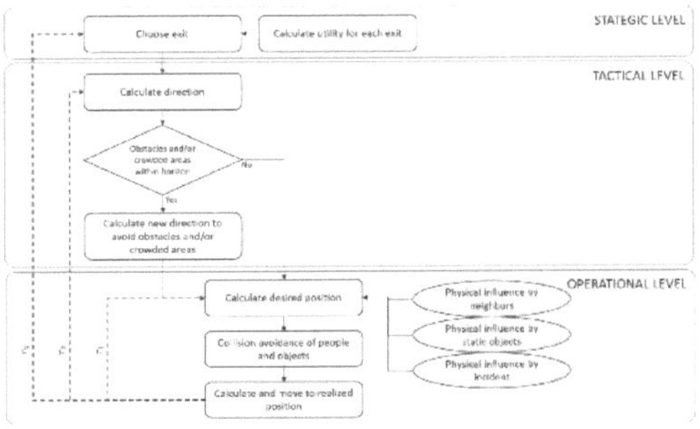

Figure 2
Three-level decision-making process of an agent.

Realistic movement under high-density conditions is simulated by collision avoidance, agents adjust their path if they detect that other agents are too close.

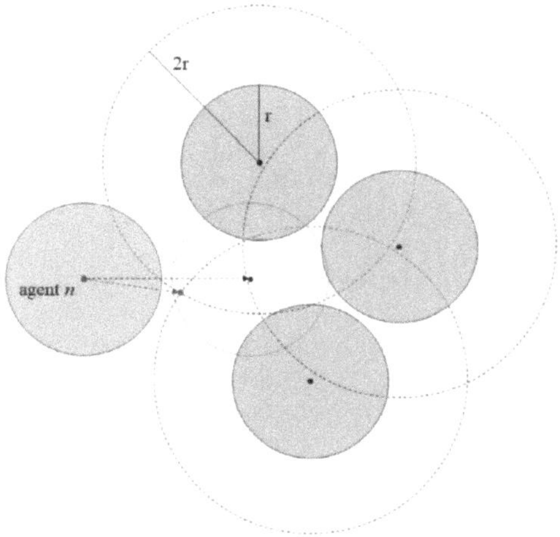

Figure 3
Collision avoidance between agents at the operational level

Utility-Based Exit Choice

Agents use choose exits using a utility exit function:

$$U_{nit}=\theta_{n,d}*D_{nit}+\theta_{n,\rho}*\rho_{it}+\theta_{n,\alpha}*\alpha_{nit}+\theta_{n,o}*o_{nit}+\theta_{n,f}*f_{nit}+\theta_{n,\epsilon}*\epsilon_{nit} \quad (1)$$

Variables:

- D_{nit}: distance from agent n to exit i at time t.
- ρ_{it}: crowd density near exit i at time t.
- α_{nit}: angle between the agent's position and the incident location as seen from exit i.
- o_{nit}: obstacle indicator (1 if the path from agent n to exit i is blocked by an obstacle; 0 otherwise).
- f_{nit}: fraction of neighboring agents also choosing exit i as their strategic goal.
- ϵ_{ni}t: random error term to introduce variability in behavior.

Parameters:

$\theta_{n,d}$, $\theta_{n,\rho}$, $\theta_{n,\alpha}$, $\theta_{n,o}$, $\theta_{n,f}$, $\theta_{n,\epsilon}$, are agent specific parameters, which define the weight attributed to each individual variable. These parameters vary depending on the agent's personality type, and result in each agent's unique behavior.

CASE STUDIES

Two case studies are used to evaluate the proposed emergency exit layout: a large outdoor music concert in Trondheim, Norway and a smaller indoor L-shaped arena. These cases help testing how grid size, emergency exit width, number of emergency exits, and risk assessment impact evacuation efficiency and safety.

Large-Scale Concert Venue: Leangen Trotting Track, Trondheim

The first case study is a Rammstein concert in Trondjeim, on the 25th of July 2021, with an expected attendance of 60.000 people, which due to COVID-19 restrictions, was canceled, but it served as a real-world case to test the Emergency Exit Layout Problem.

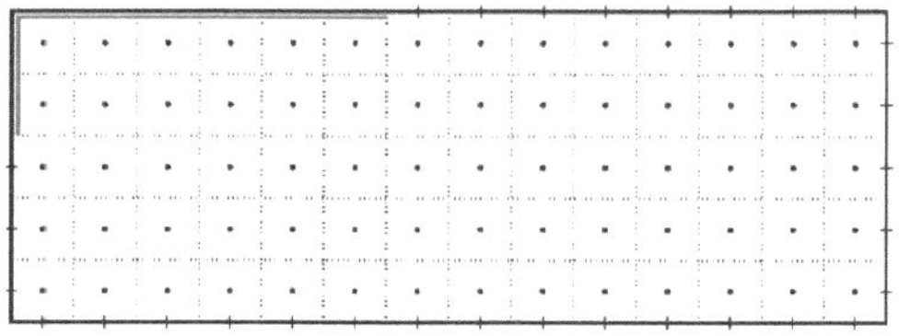

Figure 4
Arena divided into zones. Dots represent the center of the zones. Redlines represent the part of the boundary that cannot be used as emergency exits.

The area modeled is 280 by 110 meters, in a rectangular shape, divided in 3 sections, with varying crowd densities. Due to physical constraints, such as buildings and infrastructure, some parts of the boundary are unsuitable for emergency exits, and these restrictions are included in the optimization model, ensuring that exit placement respects real world limitations.

To evaluate how risk assessment impacts emergency planning, the study defines 3 separate scenarios:

- I1 – A simple scenario with 1 general evacuation, triggered by a fire alarm, with no visible hazard.
- I5 – 4 additional fire scenarios based on risks from pyrotechnic towers, which were part of the concert's stage design.
- I9 – 4 more scenarios, representing fires on food trucks, resulting in a total of 9 evacuation scenarios.

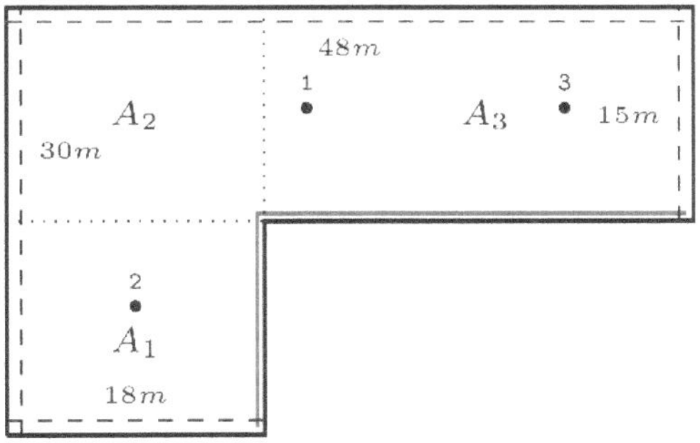

Figure 5
L-shape arena.

Each scenario is assigned probabilities, reflecting the risk assessments from local emergency services, which emphasized pyrotechnics and food trucks as the primary risks.

To make computation easier, the dimensions of the venue and the population was scaled down. Dimensions were reduced by a magnitude of 3 and the number of agents was reduced by a magnitude of 9, resulting in a 93 by 36-meter arena, and 6.700 agents.

A 3 by 3 grid (G3) was implemented, as it provided the best relation between accuracy and computational efficiency. The resulting simulations showed that the width of the emergency exits had a significant impact on the evacuation times, while the number of exits, did not consistently impact the results.

Both the optimization and simulation validated the EEPL framework, this case study also showed that in large, open areas, evacuation dynamics are primarily driven by crowed volume and exit width.

Small Indoor Arena: L-Shaped

The second space to be analyzed explores the Emergency Exit Layout Problem in a indoor L-shaped arena.

This space is non-existent and serves only for the academic purposes of this simulation. Unlike the previously analyzed case, this simulation focusses on how spatial complexity and uncertainty in agents' distribution impacts exist planning. The space consists of 3 sections, A1, A2 and A3, with a limitation on where the emergency exits can be placed, with 2 sides of the structures not suitable for emergency exits, as shown in the image below.

The agents are distributed as follows:

- D1 – even distribution of agents across A1, A2 and A3.
- D2 – Agents are concentrated on A1.
- D3 – Agents are concentrated primarily on A3.

Table 1
Information about the pedestrian distributions.

Name	A_1	A_2	A_3
D1	1/3	1/3	1/3
D2	0,7	0,2	0,1
D3	0,1	0,2	0,7

In addition, 4 incident types are defined, 1 general evacuation, and 3 specific incidents located in different areas of the venue. These incidents and distributions result in 12 different evacuation scenarios, which will be used for optimization and the simulation.

A 3 by 3 grid (G3) will be used again, resulting in 111 zones, and 37 potential emergency exit points. The goal is to choose 3 emergency exits, with a fixed width of 4 meters each. The crowd will consist of 1.500 agents, and the evacuation is considered complete when 95% of agents have exited the venue.

Four situations are defined to assess the impact of risk modeling:

- In1 – one distribution, and one general evacuation.
- In2 – all distributions, and one general evacuation.
- In3 – one distribution, and all incident types.
- In4 – all distributions, and all incidents.

The resulting location of the emergency exits show clear differences depending on the scenario, notably, In1 and In2 resulted in some zones having no reachable exits, particularly in fire scenarios, in contrast with In3 and In4, that successfully had access to all zones, indicating that detailed incident modeling significantly enhances safety in constrained environment.

In this case study it is shown that unlike in large open areas, small, complex layouts are sensitive to both agents' distribution, and agent location.

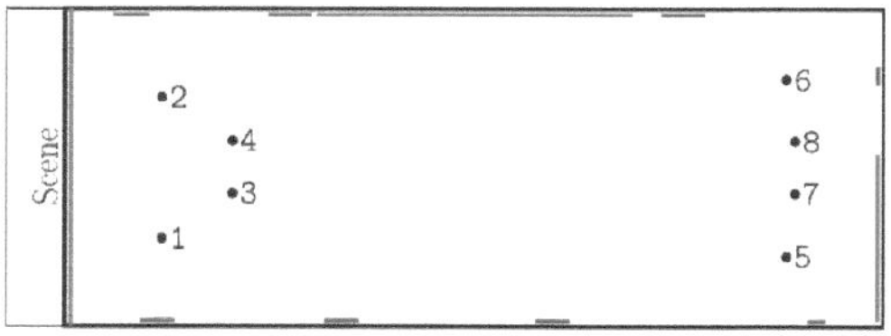

Figure 6
Layout example. Blue lines represent emergency exits.

Without detailed risk assessment, optimized layouts may leave part of the crowd unable to evacuate.

THE ALGORITHM

The goal of the algorithm is to minimize the evacuation time by choosing the optimal location and width for the emergency exits, considering incidents and crowd distribution.

Input

- Z : Set of zones in the arena.
- P : Set of possible exit points.
- T : Set of time periods.
- S : Set of scenarios.
- N : Number of exits to place.
- B: modules, each of width W, for placing at exits to calculate their width.
- W: Width of each module.
- K: Max number of pedestrians that can leave per module per time period.
- R_{is}: Number of pedestrians in zone i in scenario s before the incident.
- H: Number of pedestrians that must have left the arena for it to be considered evacuated.
- ρ_s: Probability of scenario s.

Decision Variables

- $y_p \in \{0,1\}$: 1 if an emergency exit is located at point p, 0 otherwise.
- $s_p \in \{0,\dots,B\}$: number of width modules allocated to exit p.
- $f_{ipts} \geq 0$: flow from zone i to exit p at time t in scenario s.
- $q_{pts} \geq 0$: queue at exit p at time t in scenario s.
- $e_{pts} \geq 0$: pedestrians exiting at point p, time t, scenario s.
- $w_{ts} \in \{0,1\}$: 1 if H people have evacuated by time t in scenario s, 0 otherwise.

Stage 1 – Design Decisions

$$z^T = \min \sum_{s\in\mathcal{S}} \rho_s \sum_{t\in\mathcal{T}} t w_{ts} \tag{1}$$

$$\sum_{p\in\mathcal{P}} y_p = N \tag{2}$$

$$\sum_{p\in\mathcal{P}} s_p = B \tag{3}$$

$$s_p \leq B y_p \quad p\in\mathcal{P} \tag{4}$$

$$\sum_{p\in\mathcal{P}} \sum_{t\in\mathcal{T}} f_{ipts} = R_{is} \quad i\in\mathcal{Z}, s\in\mathcal{S} \tag{5}$$

$$\sum_{t \in \mathcal{T}} f_{ipts} \leq R_{is} y_p \quad i \in \mathcal{Z}, p \in \mathcal{P}, s \in \mathcal{S} \tag{6}$$

$$\sum_{i \in \mathcal{Z}} f_{ipts} + q_{pts} = e_{pts} + q_{p,t+1,s} \quad p \in \mathcal{P}, t \in \mathcal{T}, s \in \mathcal{S} \tag{7}$$

$$e_{pts} \leq K s_p \quad p \in \mathcal{P}, t \in \mathcal{T}, s \in \mathcal{S} \tag{8}$$

$$\sum_{p \in \mathcal{P}} \sum_{\tau \in \mathcal{T} | \tau \leq t} e_{p\tau s} \geq H w_{ts} \quad t \in \mathcal{T}, s \in \mathcal{S} \tag{9}$$

$$\sum_{t \in \mathcal{T}} w_{ts} = 1 \quad s \in \mathcal{S} \tag{10}$$

$$f_{ipts} \geq 0 \quad i \in \mathcal{Z}, p \in \mathcal{P}, t \in \mathcal{T}, s \in \mathcal{S} \tag{11}$$

$$q_{pts}, e_{pts} \geq 0 \quad p \in \mathcal{P}, t \in \mathcal{T}, s \in \mathcal{S} \tag{12}$$

$$s_p \in \{0, \ldots, B\} \quad p \in \mathcal{P} \tag{13}$$

$$y_p \in \{0, 1\} \quad p \in \mathcal{P} \tag{14}$$

$$w_{ts} \in \{0, 1\} \quad t \in \mathcal{T}, s \in \mathcal{S} \tag{15}$$

$$f_{ipts} \geq 0 \qquad i \in \mathcal{Z}, s \in \mathcal{S}, p \in \mathcal{P} \setminus \mathcal{P}_{is}, t \in \mathcal{T} \mid t = \lceil D_{ip}/V \rceil \tag{16}$$

$$w_{ts} \in \{0, 1\} \qquad t \in \mathcal{T}, s \in \mathcal{S} \mid t \geq \sum_{i \in \mathcal{Z}} R_{is}/(K \cdot B) \tag{17}$$

$$z^D = \min \sum_{s \in \mathcal{S}} \rho_s \sum_{i \in \mathcal{Z}} \sum_{p \in \mathcal{P}} \sum_{t \in \mathcal{T}} D_{ip} f_{ipts} \tag{18}$$

$$\sum_{s \in \mathcal{S}} \rho_s \sum_{t \in \mathcal{T}} t w_t \leq (1 + T^r) T^E \tag{19}$$

(1) minimizes the evacuation time, weighted by probability.
(2) specifies the number of emergency exits available.
(3) specifies the number of modules available.
(4) ensures that modules are not assigned to exit points lacking exits.
(5) manages the total flow originating from each zone.
(6) guarantees that no flow is directed to exit points without exits.
(7) balances the flow passing through each exit point.
(8) sets the capacity limit for each exit.
(9) establishes the connection between the flow through exits and the number of pedestrians required to have exited the arena for evacuation to be considered complete.
(10) ensures that exactly one evacuation time is selected for every scenario.
(11) to (15) define the variables used in the model.
(16) sets to 0 all variables from constraints (11) that are not explicitly declared in (16).
(17) the total number of pedestrians in the crowd and the capacity of the exits are used to restrict w_{ts}, and equation (15) is changed to (17). The w_{ts} not declared are set to 0.
(18) places the exits so that the total probability weighted distance, but the evacuation time is not considered.
(19) limits the acceptable increase of the evacuation time.

CONCLUSION

This is a new approach emergency exit planning, integrating a two-stage stochastic optimization model with a agent based simulation framework. The Emergency Exit Layout Problem goal is to minimize evacuation time, by optimizing the number, location and width of exits under uncertain incident scenarios and agents' distributions. The two case studies demonstrate the model's effectiveness, and the results show that in large open spaces, exit width has the most impact on evacuation efficiency, while detailed risk assessment have limited influence. On the other hand, smaller and more complex venues are dependent of incident location and crowd distribution, making scenario modeling and simulation essential. The agent-based simulation helps validate the layouts and reveal behavioral dynamics not considered in optimization. These methods offer a robust tool for planners and safety officials when designing safer and more efficient evacuation strategies for mass gatherings.

REFERENCES

Barth, M. S., Palm, K., Andersson, H., Granberg, T. A., Gullhav, A. N., & Krüger, A. (2023). Emergency exit layout planning using optimization and agent-based simulation. Computational Management Science, 21(1). https://doi.org/10.1007/s10287-023-00482-y (Accessed 10 April 2025)

Li, N., Shi, Z., Jin, J., Feng, J., Zhang, A., Xie, M., Min, L., Zhao, Y., & Lei, Y. (2024). Design of Intelligent Firefighting and Smart Escape Route Planning System Based on Improved Ant Colony Algorithm. Sensors, 24(19), 6438. https://doi.org/10.3390/s24196438 (Accessed 8 April 2025)

Tsai, R.-G., Zhao, J., Yu, Y., Lin, C., Chen, X., Zheng, L., Lin, Z., & Zou, W. (2024). ERP: Real-time evacuation route planning algorithm for fire scenes in smart buildings [Preprint]. SSRN. https://ssrn.com/abstract=4748939(Accessed 17 April 2025)

Zhang, Y.-Q., Wang, J.-H., Wang, Y., Jia, Z.-C., Sun, Q., Pei, Q.-Y., & Wu, D. (2024). Intelligent planning of fire evacuation routes in buildings based on improved adaptive ant colony algorithm. Computers & Industrial Engineering, 194, 110335. https://doi.org/10.1016/j.cie.2024.110335 (Accessed 20 April 2025)

Algorithmic Methods for Automated Thermal Zoning in Building Energy Simulation

A Comparative Review of Current Approaches

Antonia Alice Hochleitner,
TU Wien,
e12223921@student.tuwien.ac.at

Thermal zoning is an essential component of building energy simulation, as it directly impacts the accuracy and applicability of simulation results. Conventional methods are still common but often time-consuming, subjective, and hard to scale. This paper reviews current algorithmic approaches for automated thermal zoning, focusing on geometry-based, rule-based, data-driven, and grid-based methods. It also discusses their use in simulation tools and the role of ontology-based frameworks in digital workflows. Finally, it outlines future perspectives, with Physics-Informed Machine Learning as a promising direction for combining data-driven methods with physical consistency.

Keywords: *Automated thermal zoning, Building energy simulation, Algorithmic methods.*

INTRODUCTION

With the increasing requirements for energy-efficient building planning, the precise simulation of thermal behavior has become a crucial component in the planning process. Thermal zoning, which is the division of a building into areas with similar heating and cooling requirements, is fundamental to generating reliable simulation results (Morkunaite et al., 2025). However, conventional approaches, such as the widely used "core and perimeter" zoning model recommended by ASHRAE 90.1, often simplify complex geometries too greatly and do not reflect the true thermal dynamics of buildings. These methods can lead to significant simulation inaccuracies (Shin and Haberl, 2022). To overcome these limitations, algorithmic methods for automatic thermal zoning are becoming increasingly important. They enable scalable, reproducible and often more precise zone definitions.

This paper provides an overview of the state of the art and a critical evaluation and comparison of current algorithmic methods for automatic thermal zoning.

CONVENTIONAL VS. AUTOMATED THERMAL ZONING STRATEGIES

Conventional zoning strategies are based on a manual or rule-based division of buildings into thermal zones. These are usually defined by planners or simulation experts. One widely used method is the "core and perimeter" model described in ASHRAE 90.1, where spaces close to the building envelope are assigned to perimeter zones and interior spaces are grouped into a core zone (Shin and Haberl, 2022). Alternatively, zones can be grouped based on the room usages, floor levels, orientation or glazing ratio. These methods are simple and appropriate for small projects, but they are often subjective, time consuming and difficult to apply to multiple buildings (Dogan et al., 2015).

Automated zoning strategies, on the other hand, use algorithmic approaches to divide buildings into thermal zones. This is done either on the basis of geometry, standardized rules or available building data. Examples of such methods include geometry-based partitioning, rule-based algorithms like Autozoner, data-driven clustering based on measured values, and grid-based zoning approaches. The main goal is to reduce manual effort, improve consistency, and make zoning scalable for complex or large-scale projects. Depending on the type of approach, automated zoning can also offer better alignment with thermal behavior, especially in early design phases or when dealing with many building models simultaneously (Dogan et al., 2015; Xiang et al., 2024).

The following table provides a comparative overview of conventional and automated thermal zoning approaches and their characteristics.

	Conventional	**Automated**
Zoning Basis	Experience, Standard Rules	Geometry, Rules, Sensor Data
Example Methods	Core/Perimeter Model, Room Usage, Orientation	Convex Partitioning, Autozoner, PCA + Clustering
Modelling Effort	High	Lower
Subjectivity	Subjective	Objective and Reproducible
Scalability	Limited	Very high
Accuracy	Moderate to Low	Moderate to High

Table 1 Comparison of Conventional and Automated Thermal Zoning Methods

The automated methods are most efficient for early-stage planning and modeling at the urban scale. However, they can be associated with higher computational requirements, a dependency on data availability and limited integration with standard simulation tools such as EnergyPlus or Rhino Grasshopper Honeybee (Dogan et al., 2015; Chang and Hsieh, 2025).

OVERVIEW OF RECENT ALGORITHMIC METHODS

Various algorithmic approaches have been developed to automate thermal zoning in building energy simulations. These include geometry-based methods, rule-based algorithms, data-driven clustering using sensor data, grid-based clustering techniques, and ontology-based data models. Each of these differs in methodology, data requirements, and intended applications.

An example of a geometry-based approach is the Convex Partition Zoner (CPZ), which divides a building floor plan into convex regions and then applies a wavefront propagation algorithm to generate zones. This method avoids the geometric complications of concave zones and enables high modeling precision. Therefore, it can be applied in early design scenarios without the need for HVAC data (Xiang et al., 2024).

Rule-based approaches such as the Autozoner algorithm automate zoning standards such as ASHRAE 90.1 by implementing computational geometry methods. This enables standardized, reproducible zone layouts and is particularly useful when applied to simplified or parametric models (Dogan et al., 2015).

Another strategy uses data-driven methods which are based on actual building data. Thermal similar rooms can be identified using time series of e.g. temperature or CO_2 sensors. One such approach uses principal component analysis (PCA) followed by k-means clustering to group spaces based on their thermal behavior. This method offers high accuracy but depends on sensor data and is more suitable for later design stages (Morkunaite et al., 2025).

An additional method is grid-based clustering, where the building is first divided into grid units. These are grouped into zones based on thermal similarity, such as correlation of temperature profiles. This method allows for detailed control and is scalable, but relies heavily on simulation data quality (Shin and Haberl, 2022).

Finally, ontology-based data integration frameworks such as the Foundation Data Model (FDM) provide a semantic structure for linking geometry, system, and usage data. These models help unify heterogeneous inputs and form a basis for consistent, automated zoning, especially in Digital Twin contexts (Xie et al., 2022).

EVALUATION AND APPLICATION AREAS OF ALGORITHMIC METHODS

Each zoning approach has different strengths depending on the planning context, available data and intended level of detail. Geometry- and rule-based methods work particularly well in early design phases when detailed building data is still limited. These methods allow simplified energy models to be created quickly and are supported by common simulation tools. This makes them particularly useful for creating schematic design models or for analyzing energy performance on a larger urban scale (Xiang et al., 2024).

Geometry-based approaches, such as the Convex Partition Zoner (CPZ), are known for their speed and robustness. They generate valid thermal zones without HVAC data. Although they can have difficulties with irregular floor plans, CPZ has proven to provide simulation results that are very close to manual zoning models while significantly reducing the modeling effort. (Xiang et al., 2024)

Rule-based systems such as the Autozoner allow for standardized and repeatable design layouts. However, they are still dependent on predefined logic and do not capture local thermal interactions or differentiated spatial conditions. (Dogan et al., 2015)

Data-driven methods provide the most accurate representation of actual thermal behavior. These approaches can improve the reliability of simulation results. However, they require comprehensive sensor data and considerable computational resources, which limits their applicability in early design phases. (Morkunaite et al., 2025)

Grid-based clustering methods offer a flexible hybrid solution. They divide buildings into spatial units and group them into zones based on statistical correlations of thermal behavior. This enables high spatial resolution and better control of thermal dynamics, especially in large or complex buildings. However, the accuracy strongly depends on the selection of correlation thresholds and the quality of the input data. (Shin and Haberl, 2022)

Ontology-based data models enhance interoperability in digital workflows by ensuring consistent integration of inputs across tools and platforms. They are especially useful for complex models and Digital Twin applications (Xie et al., 2022).

In practice, the success of these zoning methods also depends on the tools used to implement them. A comparative study (Chang and Hsieh, 2025) of simulation platforms showed significant differences in the way the tools handle zoning automation, parameter control and user interaction. Some tools offer advanced options such as scripting, visual workflows or BIM integration for flexible zone modeling. Others are more limited and may not support automated approaches as efficiently.

As a future perspective, physically informed machine learning (PIML) represents a new field with great potential for thermal zoning. These methods combine physical models with machine learning to achieve high prediction accuracy with lower data requirements. While PIML has mainly been used for predicting energy loads and optimizing controls, its integration into zoning workflows could help fill the gap between data-driven accuracy and physical consistency. However, this area is still under research and has not yet been widely used for automated zoning (Jiang et al., 2023).

CONCLUSION AND OUTLOOK

Automated thermal zoning is becoming increasingly important in building performance simulation. By replacing manual and often subjective practices with algorithmic approaches, it is possible to improve both the efficiency and reliability of simulation models. This paper reviewed and compared current methods for automated thermal zoning. Geometry- and rule-based methods are well suited for early design phases due to their simplicity and compatibility with tools like EnergyPlus. Data-driven and grid-based approaches offer greater accuracy by capturing real thermal dynamics, but require more data and computational effort. Ontology-based frameworks can further support zoning in Digital Twin environments. Looking ahead, Physics-Informed Machine Learning could help combine physical accuracy with the flexibility of data-driven methods.

Further work should focus on integrating these methods into real design processes, testing them across different building types, and developing clear criteria for evaluating their performance. The choice of approach should always reflect the specific context including the project's goals, available data, and the level of detail required.

REFERENCES

Morkunaite, L., Rasheed, A., Pupeikis, D., Angelakis, V. and Davidsson, T. (2025) A data-driven building thermal zoning algorithm for digital twin-enabled advanced control. Energy & Buildings, 336, 115633. Available at: https://doi.org/10.1016/j.enbuild.2024.11393 (Accessed 21 May 2025)

Shin, M. and Haberl, J.S. (2022) A procedure for automating thermal zoning for building energy simulation. Journal of Building Engineering, 46, 103780. Available at: https://doi.org/10.1016/j.jobe.2021.103780 (Accessed 21 May 2025)

Xiang, J., Dang, Q., Cerezo Davila, C. and Samuelson, H. (2024) Convex Partition Zoner: A New Algorithm for Automated Thermal Zoning. SimBuild 2024, IBPSA-USA. Available at: https://publications.ibpsa.org/proceedings/simbuild/2024/papers/simbuild2024_2210.pdf (Accessed 22 May 2025)

Dogan, T., Reinhart, C.F. and Michalatos, P. (2015) Autozoner: An algorithm for automatic thermal zoning of buildings with unknown interior space definitions. Journal of Building Performance Simulation. Available at: https://doi.org/10.1080/19401493.2015.1006527 (Accessed 22 May 2025)

Xie, X., Moretti, N., Merino, J., Chang, J., Pauwels, P. and Parlikad, A.K. (2022) Enabling Building Digital Twin: Ontology-Based Information Management Framework for Multi-source Data Integration. IOP Conference Series: Earth and Environmental Science. Available at: https://doi.org/10.1088/1755-1315/1101/9/092010 (Accessed 22 May 2025)

Chang, J.C., Chang, Y.T., and Hsieh, S.H. (2025) Building Energy Simulation Tools for Parametric Modeling of Building Archetypes: A Comparative Study. CAADRIA 2025 Conference Paper. Available at: https://www.researchgate.net/publication/391275488 (Accessed 22 May 2025)

Jiang, Z., Wang, X., Li, H., Hong, T., You, F., Drgona, J., Vrabie, D. and Dong, B. (2023) Physics-informed machine learning for building performance simulation – A review of a nascent field. Available at: https://doi.org/10.48550/arXiv.2504.00937 (Accessed 22 May 2025)

Beyond the Envelope: Passive Building Strategies under Urban Heat Island Conditions

A Simulation-Based State-of-the-Art Review

Elisabeth Großberger[1]
[1]*TU Wien*
[1]*e52006960@tuwien.ac.at*

This paper reviews recent simulation-based studies of passive heat mitigation in buildings under Urban Heat Island (UHI) conditions. Focusing on tools like EnergyPlus, IDA-ICE, and parametric workflows, it evaluates how boundary conditions, user behavior, and climate scenarios are represented. While some approaches use optimization and future weather data, key gaps remain—particularly in modelling microclimate variation and behavioral realism. The review calls for more context-aware, algorithmically driven simulations to support resilient design in urban heat contexts.

Keywords: *Building Performance, Climate-Responsive Architecture, Urban Heat Island*

INTRODUCTION

Rising global temperatures and increasing urban density have intensified heat stress in cities. The Urban Heat Island (UHI) effect, where urban areas remain significantly warmer than their surroundings, amplifies overheating risks, particularly during heatwaves (Kumar et al., 2024). This affects not only the building performance but also the public health, with indoor environments becoming less comfortable and more difficult to keep cool (Barone et al., 2021).

Passive design strategies—for example natural ventilation and shading—are important for reducing energy use and improving resilience (Lauss et al., 2024). In architectural terms, resilience specifies to a building's capacity to maintain critical indoor conditions by withstanding, and recovering from environmental stressors such as extreme heat (Bulman, 2024). However, the effectiveness of such strategies is highly dependent on the local outdoor conditions (Zhang et al., 2024). Despite this, building performance simulations (BPS) often rely on generic data, overlooking negative effects caused by UHI (Moradi et al., 2023).

This paper reviews passive heat mitigation strategies at the building scale, with a focus on their performance under UHI-modified boundary conditions. It draws on recent simulation-based studies that utilize engines such as EnergyPlus, often in combination with interfaces like Grasshopper, and dedicated building performance tools such as IDA-ICE. Key metrics, including Overheating Degree Hours (ODH) and Exceedance Frequency (EF), are evaluated in terms of relevance to climate resilience (Baba et al., 2023; Villaverde et al., 2023).

The aim is to assess how UHI alters passive design outcomes and how current simulation frameworks account for urban thermal conditions. The review supports the development of more accurate, context-sensitive design practices for overheating mitigation in a warming urban climate.

Urban Microclimate as Boundary Condition

Although the influence of Urban Heat Island (UHI) effects is increasingly recognized in building performance analysis, their integration into simulation workflows remains limited. Most simulations are based on standard weather datasets like IWEC (International Weather for Energy Calculations) or TMY (Typical Meteorological Year), which often rely on historical data from the 1980s to early 2000s and collected at peri-urban sites such as airports (ASHRAE, n.d.; Moradi et al., 2023). These datasets tend to underrepresent current urban heat intensities and overlook local microclimate variation, resulting in a systematic underestimation of overheating risk. A review by Li et al. (2019), spanning 24 case studies across Europe and North America, found that accounting for UHI can increase simulated cooling demand by a median of 19% (ranging from 10–120%) and reduce heating loads by 18.7% (3–45%) compared to conventional models (Li et al., 2019).

To overcome these limitations, researchers have begun incorporating UHI-adjusted conditions employing different strategies. One common approach involves modifying historical weather files through statistical morphing to reflect future urban climates. Others use spatially explicit simulations that account for urban geometry, surface properties, and materiality (Lauss et al., 2024; Pont et al., 2024; Zhang et al., 2024). Additionally, tools such as the Urban Weather Generator (UWG), available as a Ladybug component, can generate urban-modified weather files, while software tools like ENVI-met, a microscale climate model, enables detailed simulation of site-specific thermal interactions. Both tools offer additional options for assessing Urban Heat Island (UHI) effects in the context of building performance (Bueno et al., 2013; ENVI-met, n.d.). These methodologies are beyond the scope of this paper but highlight the growing effort to align simulation inputs with urban realities. Accounting for urban microclimates as dynamic boundary conditions is crucial for realistic evaluation of passive strategies, especially in dense urban areas where cooling demand is high.

PASSIVE STRATEGIES AND EVALUATIONS AT BUILDING LEVEL

Passive design strategies aim to improve thermal comfort and reduce reliance on mechanical cooling by leveraging architectural elements and natural energy flows (Archiroots, 2023). Their effectiveness is highly context-dependent, especially in urban environments where microclimate conditions, including elevated nighttime temperatures and reduced air movement due to UHI, can significantly alter performance outcomes (Li et al., 2019). This section reviews key passive strategies commonly applied at the building scale and evaluates their resilience under heat-stressed urban conditions, drawing from simulation-based studies. The strategies discussed were selected for their prominence in recent simulation-based studies.

Shading and Solar Control

External shading systems reduce solar gains by blocking direct and diffuse radiation before they enter the building envelope. Their effectiveness depends mainly on façade orientation and urban context. South-facing façades are the most straightforward to shade effectively: during summer, the high solar altitude allows horizontal overhangs or fixed louvers to block radiation while preserving daylight and winter gains (Lauss et al., 2024). East and west façades present a greater challenge. Low-angle sunlight during morning and evening hours is harder to block and often responsible for peak overheating, especially in zones with large glazing (Baba et al., 2023).

The study by Lauss et al. (2024) uses TRY2045x—a morphing-based future-focused weather file reflecting intensified summer conditions in urban Germany—as input for simulating mid-century performance. Under these scenarios, external sun shading proves its importance: when combined with triple glazing, heavy construction, and night ventilation, it enabled up to 35% energy savings and helped maintain exceedance frequencies below 1–3% in most room types. (Lauss et al., 2024). In a complementary approach, Kyogoku and Takebayashi (2024) evaluated retroreflective window films applied to east and west façades and found reductions in thermal heat gains without increasing outdoor thermal loads (Kyogoku & Takebayashi, 2024). While differing in method—geometry versus materiality—both studies emphasize the critical role of façade-level solar control in mitigating heat stress and increasing resilience. However, in dense urban contexts, reflected solar radiation from adjacent buildings can still penetrate shaded building and outdoor areas, diminishing overall effectiveness (Pont et al., 2024). As such, shading design must consider not only solar angles but also urban geometry and unwanted reflectance.

Natural Ventilation and Night Cooling

Natural ventilation, particularly night flushing, is a widely used passive strategy to remove internal heat and reduce cooling demand. It relies on outdoor air being cooler than the indoor environment and on sufficient pressure differentials to drive airflow. However, under UHI conditions, where nighttime urban temperatures often remain elevated, this cooling potential is significantly reduced. As a result, buildings in dense urban areas may retain accumulated heat, particularly when cross-ventilation is limited by surrounding obstructions or poor air exchange (Li et al., 2019; venturewell, n.d.).

Simulation studies have demonstrated this vulnerability. Baba et al. (2023) found that in a naturally ventilated Canadian school, night cooling became ineffective during periods with sustained nighttime temperatures above 24 °C, a threshold increasingly common in UHI-affected urban settings. Similarly, Lauss et al. (2024) reported that under TRY2045x conditions, the benefits of night ventilation were largely diminished, especially in lightweight constructions (Baba et al., 2023; Lauss et al., 2024).

The effectiveness of ventilation further depends on occupants' habits and window control strategies. Villaverde et al. (2023) showed that automated, climate-responsive window openings performed significantly better than fixed schedules, reducing peak indoor temperatures and overheating hours in Ladybug simulations. However, their model did not explicitly account for urban heat island effects. Overall, while night ventilation remains a low-energy cooling strategy, its resilience under UHI-intensified climates is limited.

Internal Load Reduction and Adaptive Behavior

In dense urban environments, internal heat gains from appliances, lighting, and occupant presence can significantly contribute to indoor overheating—especially when other passive strategies are constrained by UHI conditions. While internal gains are a critical factor in thermal performance, they are often kept constant in simulation studies (Peacock et al., 2010).

Peacock et al. (2010) emphasizes that high internal loads significantly increase the frequency and severity of overheating in naturally ventilated dwellings. Their work pointed out that simulation-based comfort assessments can overestimate the effectiveness of passive strategies when internal gains are not carefully calibrated. They also noted that occupant behavior—such as window opening—is frequently idealized in models, overlooking real-world constraints like noise, air pollution, or safety concerns that are especially relevant under UHI conditions (Peacock et al., 2010).

In contrast, Villaverde et al. (2023) explored a range of behavioral adaptations during a real heatwave, including window and shading operation. However, internal gains were held constant across all scenarios (Villaverde et al., 2023). This omission reflects a broader modelling gap: occupant-driven or operational strategies are rarely considered in simulations.

Passive strategies show potential, but their effectiveness under UHI is limited by urban form, boundary conditions, and occupant behavior. The next section examines how simulation tools and climate data capture—or overlook—these constraints.

SIMULATION METHODS AND GAPS

This section focuses on how resent simulations workflows model passive strategies under urban heat stress, with attention to the tools, algorithmic framework, and methodological gaps that influence the reliability of results.

Simulation Engines and Platforms

Two main simulation engines are employed in the studies reviewed. In the case of Lauss et al. (2024), IDA-ICE 4.8 was used to conduct dynamic thermal simulations on an office building. The study evaluated the performance of passive strategies across a matrix of 30 simulation configurations (Lauss et al., 2024). Within IDA-ICE, the building is divided into thermal zones, each assigned specific properties including internal loads, control settings, and ventilation parameters. You define the construction and couple inputs with a weather file, in this case the TRY2045x scenario (EQUA, n.d.).

In contrast, EnergyPlus formed the core simulation engine for Zhang et al. (2024), Baba et al. (2023), and Villaverde et al. (2023), implemented through Ladybug Tools in a Grasshopper/Rhino environment. Zhang et al. (2024) extended this pipeline with machine learning, applying SHAP (SHapley Additive exPlanations) to rank the influence of design variables (e.g. glazing, orientation) on overheating risk (Zhang et al., 2024).

Parametric and Algorithmic Workflows

Beyond the simulation engines themselves, the structure of the modelling workflow significantly shapes the depth of analysis. Parametric modelling and algorithmic optimization routines allow to systematically explore design permutations and stress-test passive measures. Villaverde et al. (2023) applied a manual parametric approach using Ladybug Tools to test a set of discrete scenarios during a real heatwave. While this allowed scenario comparison across behavioral and envelope adaptations, it lacked algorithmic iteration or automated exploration of design space. The process was driven by predefined assumptions, offering breadth but not optimization (Villaverde et al., 2023).

Zhang et al. (2024) advanced beyond manual scenario testing by integrating EnergyPlus into an optimization workflow that used machine learning and evolutionary algorithms to guide design selection. Their approach enabled automated evaluation of numerous design configurations and quantified the relative importance of variables influencing thermal performance. A non-dominated sorting genetic algorithm (NSGA-II) was applied to balance energy use and thermal comfort objectives, while Bayesian optimization supported efficient hyperparameter tuning (Zhang et al., 2024).

In contrast, Lauss et al. (2024) did not employ a parametric or algorithmic setup. Their workflow was based on a fixed matrix of scenarios manually assembled and interpreted (Lauss et al., 2024).

These differences underscore an important shift: while all reviewed studies rely on dynamic thermal simulation, only some embed algorithm-driven workflows to automate the actual design evaluation.

Performance Metrics under UHI conditions

Overheating risk in the reviewed studies is typically assessed using metrics such as Overheating Degree Hours (ODH) and

Exceedance Frequency (EF), which quantify the number of hours indoor temperatures exceed defined comfort thresholds (Lauss et al., 2024).

To assess comfort for temperate European climates, Lauss et al. (2024) employed the adaptive comfort model defined in DIN EN 16798-1, allowing comfort thresholds to vary with outdoor conditions (Lauss et al., 2024).

In contrast, Zhang et al. (2024) evaluated performance through a machine learning-supported workflow. While this enabled detailed sensitivity analysis, the underlying behavioral was based on scripted control logic and did not reflect the variability of real-world user behavior under UHI conditions (Zhang et al., 2024).

Similarly, Baba et al. (2023) applied fixed assumptions for occupant behavior, such as static window operation schedules and constant internal loads (Baba et al., 2023). These simplifications limit the realism of performance metrics under UHI conditions, as they overlook the situational limitations that can significantly affect the viability of passive cooling strategies.

Methodological Gaps and UHI Representation

Although the Urban Heat Island effect is widely acknowledged as a critical factor for building performance, current simulation workflows often fall short in representing its effects. While Lauss et al. (2024) incorporated TRY2045x files reflecting intensified summer conditions, these datasets represent city-level averages and do not capture microclimatic variations driven by urban geometry, surface materials, or wind flow (Lauss et al., 2024). Similarly, Zhang et al. (2024) included descriptors such as canyon aspect ratio and albedo in their parameter set, but these parameters were treated as static inputs and not dynamically linked to evolving boundary conditions or climate data (Zhang et al., 2024). None of the reviewed studies integrated microscale urban climate models—such as ENVI-met or the Urban Weather Generator (UWG)—into their simulation workflows. A further methodological gap lies in the absence of scenario uncertainty. While Zhang et al. (2024) accounted for uncertainty in optimization constraints, other studies used deterministic inputs for occupancy, control schedules, and weather, which restricts the robustness of their findings under future climate variability (Baba et al., 2023; Lauss et al., 2024). Taken together, these gaps underscore the need for more context-sensitive, adaptive modelling frameworks that better reflect urban microclimates and occupant realities.

CONCLUSION

This review has shown that while passive design strategies offer potential for mitigating overheating, their real-world effectiveness in urban environments remains highly context dependent. Most simulation workflows still rely on generic climate data and simplified behavioral assumptions, limiting their capacity to capture the impacts of Urban Heat Island effects. Although studies like Zhang et al. (2024) and Lauss et al. (2024) represent methodological advances—through optimization and climate-adjusted weather files—critical gaps persist. These include the absence of microclimate considerations and limited behavior realism. Addressing these gaps through more adaptive, site-sensitive modelling frameworks is essential for designing resilient buildings in increasingly heat-stressed urban contexts.

REFERENCES

Archiroots. (2023, August 18). *8 Powerful Passive Design Strategies And How They Are Proving Helpful.* https://archiroots.com/powerful-passive-design-strategies/

ASHRAE. (n.d.). *ASHRAE International Weather Files for Energy Calculations 2.0 (IWEC2).* Retrieved June 6, 2025, from https://www.ashrae.org/technical-resources/bookstore/ashrae-international-

weather-files-for-energy-calculations-2-0-iwec2
Baba, F., Ge, H., Wang, L., Zmeureanu, R., & Qi, D. (2023, September 4). *Passive adaptation strategies to mitigate the overheating risk in an existing Canadian school*. 2023 Building Simulation Conference. https://doi.org/10.26868/25222708.2023.1653
Barone, F., Merlier, L., Chasles, V., & Kuznik, F. (2021, September 1). *Contribution of building energy simulations to the assessment of overheating health risks in urban dwellings*. 2021 Building Simulation Conference. https://doi.org/10.26868/25222708.2021.30588
Bueno, B., Norford, L., Hidalgo, J., & Pigeon, G. (2013). The urban weather generator. *Journal of Building Performance Simulation*, *6*(4), 269–281. https://doi.org/10.1080/19401493.2012.718797
Bulman, J. (2024). *What is Resilient Architecture?* https://neumannmonson.com/blog/what-is-resilient-architecture
ENVI-met. (n.d.). *A holistic Microclimate Modelling System [A holistic microclimate model]*. Retrieved June 9, 2025, from https://envi-met.info/doku.php?id=root:start
EQUA. (n.d.). *IDA ICE - Simulation Software*. Retrieved June 7, 2025, from https://www.equa.se/en/ida-ice
Kumar, P., Debele, S. E., Khalili, S., & Jones, L. (2024). Urban heat mitigation by green and blue infrastructure: Drivers, effectiveness, and future needs. *The Innovation*, *5*(2), 100588. https://doi.org/10.1016/j.xinn.2024.100588
Kyogoku, S., & Takebayashi, H. (2024). *Study on thermal environment planning to social implementation of heat control technology in outdoor space*.
Lauss, L., Hepf, C., Schmid, T., Romero, J. A., & Auer, T. (2024, September 23). *Simulation-based evaluation of thermal comfort in buildings, by mitigation of heat stress through passive measures*. BauSim Conference 2024. https://doi.org/10.26868/29761662.2024.57
Li, X., Zhou, Y., Yu, S., Jia, G., Li, H., & Li, W. (2019). Urban heat island impacts on building energy consumption: A review of approaches and findings. *Energy*, *174*, 407–419. https://doi.org/10.1016/j.energy.2019.02.183
Moradi, A., Kavgic, M., Costanzo, V., & Evola, G. (2023). Impact of typical and actual weather years on the energy simulation of buildings with different construction features and under different climates. *Energy*, *270*, 126875. https://doi.org/10.1016/j.energy.2023.126875
Peacock, A. D., Jenkins, D. P., & Kane, D. (2010). Investigating the potential of overheating in UK dwellings as a consequence of extant climate change. *Energy Policy*, *38*(7), 3277–3288. https://doi.org/10.1016/j.enpol.2010.01.021
Pont, U., Wölzl, M., Stiegler, V., Wolffhardt, R., Schober, P., Swoboda, S., & Bauer, P. (2024, September 23). *Performative Evidence of Smart and Urban Trees—Assessing the thermal and visual impact of large scale urban shading structures via simulation*. BauSim Conference 2024. https://doi.org/10.26868/29761662.2024.32
venturewell. (n.d.). *Night-Purge Ventilation | Sustainability Workshop*. Retrieved June 6, 2025, from https://sustainabilityworkshop.venturewell.org/buildings/night-purge-ventilation.html
Villaverde, A., Garmendia, L., Egiluz, Z., Briz, E., Cuadrado, J., & Rojí, E. (2023, September 4). *Measuring the effectiveness of retrofit, behavioural and transformational adaptation solutions to heat waves*. 2023 Building Simulation Conference. https://doi.org/10.26868/25222708.2023.1254
Zhang, Y., Teoh, B. K., & Zhang, L. (2024). Multi-objective optimization for energy-efficient building design considering urban heat island effects. *Applied Energy*, *376*, 124117. https://doi.org/10.1016/j.apenergy.2024.124117

Walking the Heat Wave

A theoretical framework linking pedestrian mobility and urban heat vulnerability

Lucia Sole
TU Wien
e12307576@student.tuwien.ac.at

Climate change and Urban Heat Island effect have intensified the negative impact of heatwaves on human health, making the need of mitigation strategies more and more urgent. Understanding heat exposure as a correlation between human presence and heat stress sheds light on the possibility to integrate pedestrian mobility models into microclimate simulations as promising research direction. Although the mutual influence between pedestrian behaviour and microclimate has been explored in the past decade, many pedestrian mobility models remain difficult to apply in the context of planning due to scalability and resource demands. At the same time. while vehicle mobility models have long been used in infrastructure design, a clear and simple theoretical framework around pedestrian mobility models would be beneficial to guide their use in planning intervention practices. This paper explores pedestrian mobility models, proposing a taxonomy based on their theoretical foundations and three measurable indicators: stochasticity, dynamism, and granularity. The potential and characteristics of pedestrian mobility models within the context of urban microclimate simulations are critically discussed in relation to emerging trends and case studies.

Keywords: *Urban microclimate, Pedestrian mobility, Urban network analysis, Heat exposure, Outdoor thermal comfort.*

INTRODUCTION

The necessity to plan and prioritize urban heat vulnerability mitigation strategies, in the context of climate change, has made microclimate simulation and modelling a prominent assessment tool in the hands of planners. The relation between changing environmental conditions and vulnerability to heat stress, which has been clarified by the Intergovernmental Panel on Climate Change (*Fifth Assessment Report - Synthesis Report*, n.d.), has made the integration of pedestrian mobility models in urban microclimate simulations a growing trend. The concepts of heat stress and pedestrian occupancy are connected in the definition of heat exposure as the contact between an individual and a hostile environment, which causes health risks or discomfort due to extremely high temperatures (Kuras et al., 2017). To exhaustively estimate and map the possibility and entity of risks related to heat exposure at urban scale, said elements should be considered as a time and space variable factor, dependent on where people are and scalable in relation to the adaptability of local populations (*Heat and Health*, n.d.). More than that, there is well assessed reciprocal correlation between Urban Heat Island effect and people's behavioural and mobility patterns: while anthropogenic emissions have an impact on the environmental conditions of the cities (Oke,

1982), the presence of pedestrians in outdoor spaces and their choices regarding routes and means of transportation are influenced by the thermal environment (Nikolopoulou et al., 2001a). The feedback loop may worsen the phenomena, for example by increasing the cooling demand and the time spent in indoor spaces (Ashtiani et al., 2014), or by encouraging the use of vehicles. Thus, mobility models have a twofold use in urban microclimate simulations, helping to assess and map the risks related to heat stress and making the prediction of anthropogenic emissions in such models more granular. The following paper is focused on explaining what a pedestrian mobility model is, noting what characteristics different kinds of model present, with the aim of producing a taxonomy and underlining the possibility of integrating them in urban microclimate simulations.

PEDESTRIAN MOBILITY MODELS

In the field of animal ecology and management, gathering data to investigate population numbers and distribution is an established practice (MacKenzie et al., 2018), with the emergence of some related fields, such as movement ecology, which is specifically interested in defining the causes and mechanisms behind the behaviour of individuals in motion (Nathan et al., 2009). In this context the concept of "model" can be understood as a representation of an observable system which can be used to describe its behaviour and predict its response at the occurrence of certain perturbance scenarios (MacKenzie et al., 2018). A similar modelling definition can be applied to pedestrian mobility, which acquired a certain interest in urbanism for the pressing need to asses and improve the environmental, social and economic sustainability conditions of urban contexts (Berger et al., 2014). During the last century, the enthusiasm and necessity in assisting the development of public and private transport infrastructure, has made so that a large number of approaches have been carried out to evaluate and simulate motorized transportation trips (Ortúzar & Willumsen, 2011), but the resolution and the scale of such models is mostly incompatible with the peculiarities of pedestrian trips. The concept of walkability has gained more and more importance, as planning practices are shifting from car-centred towards pedestrian-centred city models (Sevtsuk & Kalvo, 2025). In order to highlight some key peculiarities of different kinds of models, the methodology chosen to simulate the behaviour of individuals must be distinguished from the choice of internal and external factors governing the movement. The way in which a population of individuals is modelled plays a crucial role, because it characterizes the model's granularity, dynamism and stochasticity. At the same time, models based on very different simulation approaches show to consider similar factors as the final goal for pedestrian movement. For this reason, the discussion about what drives pedestrian movement will be left out for now.

What differentiates pedestrian mobility models

While presenting the Urban Network Analysis (UNA) tools, a framework for the analysis of pedestrian urban networks, Sevtsuk & Kalvo, (2025) identified three broad categories of pedestrian mobility models:

- Direct demand models
- Agent-based models
- Network flow models

Direct demand models are based on the characteristics of the site, i.e. on sets of available site-specific data and their analysis, but without considering the roles of individuals or setting movement rules to estimate a network of possible routes; they calculate the total or average amount of pedestrian activity expected at a specific point in time, without considering the probability and quality of individual trips.

Agent-based models are focused on rule-based representations of individual-level and inter-individual level dynamics and will be discussed in a more defined context later on.
Finally, network flow models, on which the UNA framework itself is based, evaluate network qualities (such as accessibility, centrality, or potentiality) using mostly graph-based analysis, therefore mathematically. The presented classification suggests that there are some key qualities for identifying different approaches to pedestrian mobility modelling, but it is not generic enough to give a more comprehensive and easier to handle overview. What seems crucial for now is the level of determinism and extent to which models define individual agents, account for their interactions, and represent their relationship with the surrounding context. These factors can be captured by three main indicators that help recognise different approaches to mobility modelling: granularity, dynamism and stochasticity.

- Granularity measures the level of definition of individuals in a model; the question here is whether and how a model is considering the characterization of the individuals in a way that influences the analysis of possible trips and routes, and then the estimations of volume and distribution of people. The level of granularity also correlates to the distinction around microscopic or macroscopic models (Kouskoulis & Antoniou, 2017).
- Dynamism measures how much a model responds and adapts to changes over time; this means that it is important to point out if the system is reactive to time dependent environmental conditions, i.e. if it evolves based on real time feedback.
- Stochasticity measures how much probability theory and randomness are influencing the outcome of the model analysis. It is mostly related to how much the variability of human behaviour is considered in the model and how it is represented. A model will be highly deterministic if a set of given inputs always leads to the same output.

The balancing of these dimensions depends on how the model is conceived and what its foundations are, meaning the set of available data, ideas and underlying principles that guide its development.

Theoretical or Data-Driven foundation: definition and examples

A common distinction linked to the models' foundation is made on whether they are data-driven or theoretical: this seems to give an important direction to the whole model conception and to have a sensible influence on the measure of the key indicators mentioned above. Theoretical models are interested in capturing behaviours and reflecting real-world systems by defining sets of mathematical or physical rules; they are based on assumptions to capture the nature of interactions between individuals and between individuals and context. Examples of theoretical approaches are social force, cellular automata and lattice gas models, exhaustively described by Kouskoulis & Antoniou (2017), and often used as a foundation for agent-based interactions. Indeed, for what has been said, agent-based models can now be better understood as detailed representation of an environment and of the interactions in it, based on self-consistent unities determined starting from behavioural theory and case specific studies; they are granular by definition, dynamic as they evolve through interactions and often stochastic as they rely on probability to enhance variance and simulate decision-driven actions by the individuals (Kaziyeva et al., 2023). Agent-based models are not so commonly used in large scale pedestrian mobility analysis, because of their complexity, the huge cost in terms of time and resources and for the fact that small-scale interactions are less significant at larger scales. In the context of urban microclimate and outdoor thermal comfort studies, they are mostly focused on the effects that the environment has on pedestrian behaviour and vice-versa in small temporal windows (Muslimin & Brasier, 2024). With the goal of extending a pedestrian agent-based mobility model to a city scale, Kaziyeva et al. (2023) had to omit to model interactivity between people, while considering a decision-making process based on pedestrian desired activities and related routes calculations. The study was carried out in Salzburg and required the acquisition of a large amount of data from population census and surveys, including mobility diaries from more than forty thousand respondents.

Data-driven models do not require any a priori knowledge of the relationships between the system's parameters, and rely on sets of data to extrapolate trends and patterns. Depending on the data and methodology used, data-driven models can vary in stochasticity, but they are typically less granular than the theoretical ones, as they move away from individual-level behaviour to identify aggregate patterns. They may not be considered dynamic, as they reflect the space and time conditions of the datasets they are based on, rather than actively respond to real-time changes.

Data-driven pedestrian mobility models are a growing tendency, as location data have been widely digitalized through the use of tracking technologies integrated into users' smartphones (Gutiérrez-Roig et al., 2015), internet connected web mapping platforms, geo-localized social reviews (Li et al., 2024), computer vision devices (Wang & He, 2023) and more. These technologies, however indiscreet and in some cases insidious they may seem, have spread worldwide and uncovered a wide, high resolution network of data on pedestrian trips (Noulas et al., 2012). In the context of environmental urban studies, risk assessment, and planning support, data-driven models are particularly valuable because they enable the estimation of pedestrian density through quantitative analysis, helping to map and prioritize human-centred mitigation strategies.

As an example on how the described key indicators, i.e. granularity, dynamism, and stochasticity, and the methodological foundations can be used to characterize some pedestrian mobility models, Table 1 links them to the model types described by Sevtsuk & Kalvo, (2025).

Model Type	Granularity	Dynamism	Stochasticity	Foundation	Example Use / Application
Direct Demand	Low	Low	Low	Mostly Data-driven	Large-scale pedestrian volume estimation
Agent-Based	High	High	High	Theoretical	Small-scale detailed behavioural studies
Network Flow	Medium	Medium	Medium	Mixed	Urban network analysis and route accessibility

Table 2
Characterization of pedestrian mobility models based on three key indicators, and their methodological foundation.

PEDESTRIAN MOBILITY AND MICROCLIMATE: APPLICATIONS

The necessity to integrate pedestrian mobility models and microclimatic studies seems to be driven mainly by two broad research goals, which are the assessment and mapping of health risk related to heat stress in built environments, and the discussion on what is the influence that pedestrian mobility has on microclimate and vice-versa. The first goal is strictly related to the need of assisting the decision-making process around mitigation strategies and of creating knowledge and awareness about the entity of health-related risks due to climate change, while the second is a more explorative field whose answers have a huge implication on how the problem can be framed from a behavioural study point of view. These two objects of study are correlated: the documentation about the mutual influences between people's habits and microclimate is the ground base for justifying the need to integrate pedestrian mobility considerations while simulating urban microclimate or mapping health related risk.

In both cases, the question of what makes people move needs to be answered, but if this question is explicit while correlating environmental factors and mobility patterns, it is implicit in urban modelling, because here we want to understand how many people are in which part of the simulated land, to then perform other calculations and draw some conclusions.

The following subsections, one for each research goal, show through two case studies how this distinction implies a difference in the preferred methodological foundation and data acquisition strategy.

What makes people move? An agent-based case study

Muslimin & Brasier (2024) move from the correlation between thermal conditions and the use of public space (Nikolopoulou et al., 2001) to develop an agent based model; the aim is to document the impact of different environmental factors, such as wind, shade and temperature, as well as land functional factors, such as the spread of local economic activities, on the pedestrian walking patterns. The urban context is synthesized as a layered information model, by separating the geometrical characteristics from the functional ones; pedestrian mobility is intended in a goal-oriented fashion, following the distinction given by Lindenberg & Steg (2007) into normative, gain and hedonic goals. The normative goals are a consequence of social norms and guide people towards acting appropriately; they serve to model the interactions between agents and their way of moving through crowds. The gain goals are linked to the use or the acquisition of resources, and to the related considerations that may influence people's route choices, for example the need to reach a certain destination in less time, or to go to work. The hedonic goals are related to the desire of bettering the comfort conditions, for example walking on a shaded part of the street, or generally improving the way a person feels, for example walking where more amenities or streetlights are located. Figure shows how these information and the various analysis tools were integrated.

Mapping the heat stress: a data-driven case-study

To map the risks related to heat exposure while also taking into account pedestrian mobility at a district scale, Colaninno et al. (2025) developed a complex approach, where data-driven analysis enrich a stochastic network flow model. The network flow model evaluates the sidewalk connections of a portion of Los Angeles, returning the accessibility to five different kinds of amenities, i.e. public transport stations, schools, parks and commercial buildings. Residential buildings are considered as the origin for the trips' calculations, and various routes are computed using the number of people per residential building as a weight. The computational tools are developed in the already mentioned Urban Network Analysis framework and are based on some behavioural considerations abstracted as numerical thresholds, such as the distance a pedestrian could be willing to walk, and combined, while maintaining a certain stochasticity. The result of such analysis is a time-independent estimate of the pedestrians' volume distributed on the road network, which was then connected to a dataset about pedestrian counts in Los Angeles, to be calibrated and time variated. In this way Colaninno et al. (2025) obtained a distribution of pedestrian volume per point in time and then combined this information with a UTCI (Universal Thermal Comfort Index) calculation to assess a new heat exposure index and map it on the sidewalks. Such index is sensitive to both the number of pedestrians and the perceived temperature in a specific location; therefore, it is a useful tool for prioritizing and assessing strategies of intervention.

DISCUSSION AND CONCLUSION

The fact that heat stress is linked to individuals' physiological vulnerability makes mapping people's presence in the urban context essential to plan and prioritize heat mitigation strategies.

With the intention of promoting the integration of systematic ways to represent the distribution of pedestrians in urban environmental studies, the concept of pedestrian mobility models has been explained, and a taxonomy has been proposed. The distinction between different kinds of models has been carried out on the basis of two different methodological grounds (theoretical and data-driven), and of three key indicators (granularity, stochasticity, and dynamism). The difference in applicability of theoretical and data-driven models to large scale simulations has been discussed; while the first has been studied more in depth in the last decades, these studies are mostly linked to the field of behavioural ecology, and to the understanding of how the environment influences walking. In fact, they require huge amounts of time and resources, and they might not be worth the cost in microclimatic simulations. An in-depth study on how small-scale interactions between individuals and between individuals and context have an influence on a large scale could be beneficial to evaluate their applicability or to enhance the possibility of doing so.

Data-driven pedestrian mobility models are a growing research subject of the big data era, but their outcome is hard to validate. Open data about human behaviour in the cities remain limited and it is hard to obtain freely shared data from Information and Communication Technology companies. More than that, data acquired by researchers using sensors or computer vision often present privacy issues, while using data acquired through web mapping or tracking services is often part of a legal grey zone. Probably, a clearer legal and ethical framework is needed to support future research activity and ensure responsible use of these data sources.

For these reasons, mixing different modelling approaches to have a more precise idea of pedestrian mobility is a way to overcome these difficulties. The case study by Colaninno et al. (2025), set in Los Angeles, shows this well: the mathematical tools used for the analysis of the urban network, while grounded in graph theory, also consider the behavioural tendencies of the individuals, and use stochastic methods to reproduce the uncertainty of choice. The mathematical approach facilitates the integration of real-world data, enriching the model without causing conflict.

Behaviour and microclimate are in a feedback loop correlation, therefore, developing flexible frameworks which tend to reflect both small-scale and large-scale interactions at different steps seems promising in the context of heat vulnerability assessment.

REFERENCES

Ashtiani, A., Mirzaei, P. A., & Haghighat, F. (2014). Indoor thermal condition in urban heat island: Comparison of the artificial neural network and regression methods prediction. Energy and Buildings, 76, 597–604. https://doi.org/10.1016/j.enbuild.2014.03.018

Berger, G., Feindt ,Peter H., Holden ,Erling, & and Rubik, F. (2014). Sustainable Mobility—Challenges for a Complex Transition. Journal of Environmental Policy & Planning, 16(3), 303–320.

https://doi.org/10.1080/1523908X.2014.954077
Colaninno, N., Basu, R., Hosseini, M., Alhassan, A., Liu, L., & Sevtsuk, A. (2025). A sidewalk-level urban heat risk assessment framework using pedestrian mobility and urban microclimate modeling. Environment and Planning B: Urban Analytics and City Science, 52(5), 1071–1090. https://doi.org/10.1177/23998083241280746
Fifth Assessment Report—Synthesis Report. (n.d.). Retrieved 4 June 2025, from https://archive.ipcc.ch/report/ar5/syr/
Gutiérrez-Roig, M., Sagarra, O., Oltra, A., Palmer, J. R. B., Bartumeus, F., Diaz-Guilera, A., & Perelló, J. (2015). Active and reactive behaviour in human mobility: The influence of attraction points on pedestrians (No. arXiv:1511.03604). arXiv. https://doi.org/10.48550/arXiv.1511.03604
Heat and health. (n.d.). Retrieved 4 June 2025, from https://www.who.int/news-room/fact-sheets/detail/climate-change-heat-and-health
Kaziyeva, D., Stutz, P., Wallentin, G., & Loidl, M. (2023). Large-scale agent-based simulation model of pedestrian traffic flows. Computers, Environment and Urban Systems, 105, 102021. https://doi.org/10.1016/j.compenvurbsys.2023.102021
Kouskoulis, G., & Antoniou, C. (2017). Systematic Review of Pedestrian Simulation Models with a Focus on Emergency Situations. Transportation Research Record: Journal of the Transportation Research Board, 2604, Article 17–03140. https://trid.trb.org/View/1438254
Kuras, E. R., Richardson, M. B., Calkins, M. M., Ebi, K. L., Hess, J. J., Kintziger, K. W., Jagger, M. A., Middel, A., Scott, A. A., Spector, J. T., Uejio, C. K., Vanos, J. K., Zaitchik, B. F., Gohlke, J. M., & Hondula, D. M. (2017). Opportunities and Challenges for Personal Heat Exposure Research. Environmental Health Perspectives, 125(8), 085001. https://doi.org/10.1289/EHP556
Li, L., Hu, S., Shaw, A., & Hemphill, L. (2024). Crowdsourcing public attitudes toward local services through the lens of Google Maps reviews: An urban density-based perspective (No. arXiv:2404.13156). arXiv. https://doi.org/10.48550/arXiv.2404.13156
Lindenberg, S., & Steg, L. (2007). Normative, Gain and Hedonic Goal Frames Guiding Environmental Behavior. Journal of Social Issues, 63(1), 117–137. https://doi.org/10.1111/j.1540-4560.2007.00499.x
MacKenzie, D. I., Nichols, J. D., Royle, J. A., Pollock, K. H., Bailey, L. L., & Hines, J. E. (2018). Chapter 1—Introduction. In D. I. MacKenzie, J. D. Nichols, J. A. Royle, K. H. Pollock, L. L. Bailey, & J. E. Hines (Eds.), Occupancy Estimation and Modeling (Second Edition) (pp. 3–26). Academic Press. https://doi.org/10.1016/B978-0-12-407197-1.00002-8
Muslimin, R., & Brasier, N. (2024). An analysis of outdoor thermal comfort impact on pedestrian walking patterns at sub-hour timescales with agent-based modelling. Smart and Sustainable Built Environment, ahead-of-print(ahead-of-print). https://doi.org/10.1108/SASBE-06-2023-0158
Nathan, R., Getz, W., Revilla, E., Holyoak, M., Kadmon, R., Saltz, D., & Smouse, P. (2009). A movement ecology paradigm for unifying organismal movement research. Proceedings of the National Academy of Sciences of the United States of America, 105, 19052–19059. https://doi.org/10.1073/pnas.0800375105
Nikolopoulou, M., Baker, N., & Steemers, K. (2001a). Thermal comfort in outdoor urban spaces: Understanding the human parameter. Solar Energy, 70(3), 227–235.

https://doi.org/10.1016/S0038-092X(00)00093-1
Nikolopoulou, M., Baker, N., & Steemers, K. (2001b). Thermal comfort in outdoor urban spaces: Understanding the human parameter. Solar Energy, 70(3), 227–235. https://doi.org/10.1016/S0038-092X(00)00093-1
Noulas, A., Scellato, S., Lambiotte, R., Pontil, M., & Mascolo, C. (2012). A Tale of Many Cities: Universal Patterns in Human Urban Mobility. PLoS ONE, 7(5), e37027. https://doi.org/10.1371/journal.pone.0037027
Oke, T. (1982). The energetic basis of urban heat island. Quarterly Journal of the Royal Meteorological Society, 108, 1–24. https://doi.org/10.1002/qj.49710845502
Ortúzar, J. de D., & Willumsen, L. G. (2011). Modelling Transport. John Wiley & Sons.
Sevtsuk, A., & Kalvo, R. (2025). Modeling pedestrian activity in cities with urban network analysis. Environment and Planning B: Urban Analytics and City Science, 52(2), 377–395. https://doi.org/10.1177/23998083241261766
Wang, L., & He, W. (2023). Analysis of Community Outdoor Public Spaces Based on Computer Vision Behavior Detection Algorithm. Applied Sciences, 13(19), Article 19. https://doi.org/10.3390/app131910922

Semantic Web Technologies in the AEC Industry

Improving Interdisciplinary Collaboration through Semantic Enrichment

Isabelle Seifert[1]
[1]*TU Wien*
[1]*e12432970@student.tuwien.ac.at*

This paper explores how Semantic Web Technologies (SWTs) can enhance interoperability, automation, and cross-disciplinary collaboration in the Architecture, Engineering, and Construction (AEC) industry through their integration with Building Information Modelling (BIM). While BIM offers a structured representation of building data, it often lacks semantic depth, limiting its potential for advanced reasoning and seamless data exchange. Therefore, this paper presents three state-of-the-art research projects that demonstrate how SWTs can transform BIM into a dynamic, machine-readable system, enabling intelligent decision-making across the building lifecycle.

Keywords: *Semantic Web Technologies, Knowledge Graphs, Interoperability, Collaboration, Building Performance Analysis*

INTRODUCTION

The Architecture, Engineering, and Construction (AEC) industry is navigating a digital transformation marked by the increasing integration of Building Information Modelling (BIM). While BIM offers a structured digital representation of building data, it often falls short in delivering interoperability, cross-domain reasoning, and advanced automation due to limited semantic richness. Challenges such as overcoming the fragmentation of domain specific data and a lack of meaningful data integration persist across the project lifecycle (Pauwels et al., 2017).

Semantic Web Technologies (SWTs), developed around standards like RDF (Resource Description Framework), OWL (Web Ontology Language), and SPARQL (SPARQL Protocol and RDF Query Language), offer promising tools to overcome these issues. By enabling machines to interpret and reason with data meaningfully, SWTs enhance the intelligence and interoperability of BIM environments.

This paper explores the integration of Semantic Web Technologies in BIM through key applications such as cross-disciplinary design workflows, language model querying as well as waste quantification, revealing their transformative potential for the AEC sector.

SEMANTIC WEB TECHNOLOGIES

The Semantic Web, as defined by the W3C (2014), extends the current web by giving data well-defined meaning, allowing machines and humans to interact more effectively. In the AEC sector, characterized by complex, multidisciplinary collaboration and information-intensive workflows, this semantic structuring is especially valuable (Pauwels et al., 2016).

At the core of the Semantic Web are RDF Triples consisting of Subject, Predicate and Object as illustrated in Figure 1 on the left. They build a graph-based representation of data with semantic information and connections. For example, a wall in a BIM model may be semantically enriched with links to its material composition, thermal properties, and spatial relationships. Such semantic relationships are defined using ontology languages like RDF (Resource Description Framework) and OWL (Web Ontology Language), which provide the formal syntax and semantics needed for representing and reasoning about domain-specific knowledge (W3C, 2014). These technologies enable the structuring of different data in a machine-readable and logically consistent manner, while connecting them with semantic connections.

Rather than being limited to rigid naming conventions within a single software environment, Semantic Web Technologies allow the integration of heterogeneous data sources through shared ontologies. This enables the creation of interconnected knowledge graphs, as illustrated in Figure 1 on the right, that span multiple platforms and disciplines through several ontologies and semantic connections.

By embedding such semantic structure, BIM can evolve from a static data container into a dynamic, knowledge-based environment that supports automation, advanced analysis, and collaborative decision-making.

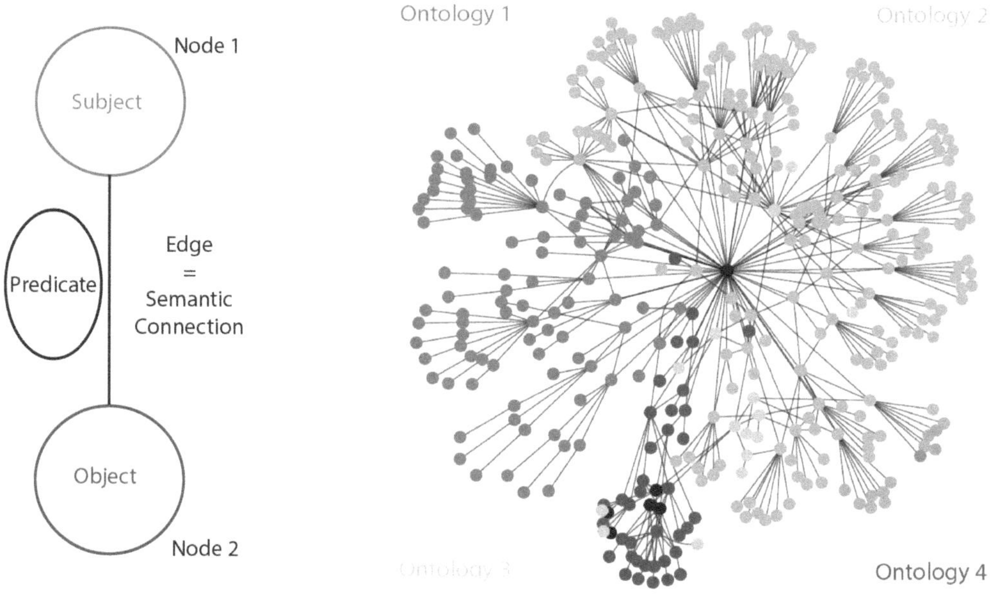

Figure 1
Semantic Web
Technologies.

STATE OF THE ART
APPLICATIONS IN THE AEC INDUSTRY

In the past years there have been numerous researchers, that investigated in semantic web technologies and their potential to close the gap between BIM Models and different databases. In the following paragraph a few state-of-the-art examples are being introduced.

Enhancement of Co-design Workflows

In their research, Elshani et al. (2025) analysed and evaluated the application of Semantic Web Technologies (SWTs) within the Architecture, Engineering, and Construction (AEC) industry, focusing specifically on enhancing co-design workflows to support cross-disciplinary collaboration. A central aim of their study was to enable seamless data exchange between different stakeholders by integrating semantically structured information across diverse platforms. To achieve this, the authors developed a mapping strategy for bidirectional translation using ontology-based formats like RDF and OWL. This approach creates not only the semantic representation of data but also enables its reconversion into domain-specific modelling environments.

BIM models, structured using the Industry Foundation Classes (IFC) standard, are particularly well-suited to this method. The IFC schema, developed by buildingSMART, inherently follows a hierarchical structure based on entities, classes, and subclass categories (buildingSMART, 2024). However, other platforms and data structures used within a construction project do not necessarily match this structure well, often leading to data loss and interoperability issues (Elshani et al. ,2025). To overcome these limitations, Elshani et al. leveraged the Buildings and Habitats Object Model (BHoM), an open-source interoperability platform. BHoM enables the creation of discipline-specific mappings, which were then semantically linked and stored within GraphDB, which is a database able to store semantic information in a semantic web structure (see Figure 2).

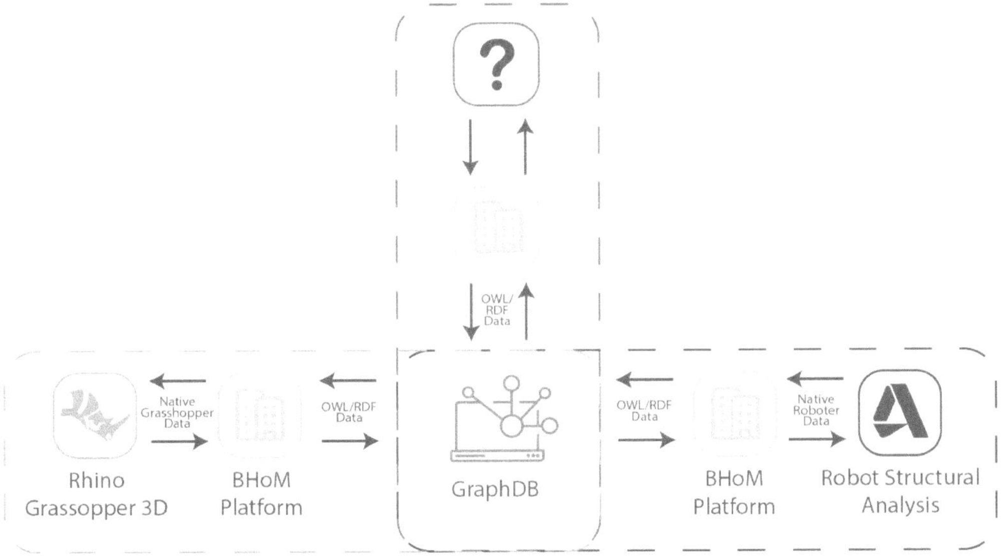

Figure 2
Co-design Workflow, adapted from Elshani et al. (2025).

The primary advantage of this approach lies in its flexibility: new data, even if originating from platforms with differing nomenclature or modelling conventions, can be integrated efficiently and added to the semantic web (Figure 2). The BHoM framework acts as a translator and semantic mediator, enabling diverse stakeholders to contribute their domain-specific data while maintaining interoperability. By connecting these various models into a unified knowledge graph, the system allows for advanced querying, reasoning, and collaboration throughout the project lifecycle.

A persisting challenge pointed out is the complexity of creating the queries and thereby extracting exact information suitable for the translation into the BHoM objects. To mitigate this issue Elshani et al. (2025) suggest the incorporation of LLMs (Large Language Models) or AI (Artificial Intelligence) to automate the mapping processes and create learning-assisted ontologies.

BIM Analysis Through Knowledge Graphs and LLMs

BIM models contain vast amounts of structured data, but querying this information often requires advanced technical knowledge. Massafra et al. (2024) address this by integrating Large Language Models (LLMs) with Knowledge Graphs derived from BIM models.

The main research question posed is:

"How can users with little technical knowledge extract the respective data from a BIM Model by only posing a prompt to an LLM"

The research approach includes the creation of a test BIM model using Topologic, a software modelling library enabling hierarchical and topological representations of architectural spaces and buildings through Non-Manifold Topology (NMT). To transform this model into a semantic graph structure, the authors used a custom-developed module called BTwin. BTwin is an open-source python library and acts as a translator of geometric and topological data. Specifically, it converts the BIM data into a graph structure using NetworkX, a Python package designed for the creation, manipulation, and analysis of complex networks (Figure 3). In this context, each building component is represented as a node, while spatial and functional relationships between components are modeled as edges. The transformation is guided by semantic principles defined in a supporting ontology, ensuring that the resulting graph captures not just geometry but also domain-specific meaning. This enables the integration of building data into a machine-readable graph that can then be queried through natural language inputs interpreted by the LLM.

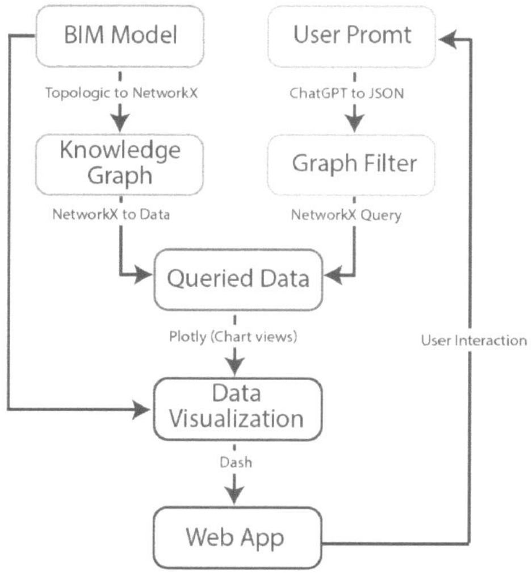

Figure 3 Research Workflow, adapted from Massafra et al. (2024).

The user interaction component of the system involves submitting a prompt, such as *"Sum the area of all office spaces"*, which is then processed by the LLM (e.g., ChatGPT) to generate a structured JSON query. This is interpreted by the Graph Filter, which translates it into a semantic query that filters the NetworkX-based knowledge graph. The resulting queried data is passed to a data visualization layer, which uses Plotly for both 3D building views and chart-based representations, all of which are rendered interactively through a Dash-based web application Massafra et al (2024) developed. The entire pipeline is illustrated in Figure 3, which shows the flow from BIM model generation and user prompt input, through graph querying and filtering, to web-based visualization and interaction.

An example of the JSON query generated through this process is shown below (Figure 4). In this case, the LLM interprets a user prompt as a task to sum the gross area (pr_GrossArea) of all spaces classified as specific office types, where each space must exceed a size of 25 square meters. The structured format includes classes, properties, filter values, and logical operators, all machine-readable and ready to be executed against the knowledge graph.

Figure 4
Example JSON graph filter.

```
{
  "Task": {
    "TaskType": "Sum",
    "PropName": "pr_GrossArea"
  },
  "Class": "Space",
  "Filter1": {
    "PropName": "pr_OccupancyType",
    "FilterValue": [
      "Office:AdministrationOffice",
      "Office:MeetingRoom",
      "Office:ResearchOffice"
    ],
    "FilterOperator": "="
  },
  "Filter2": {
    "PropName": "pr_GrossArea",
    "FilterValue": 25,
    "FilterOperator": ">"
  }
}
```

This approach allows users to access data related to spatial layouts, thermal zones, and environmental parameters without specialized knowledge of BIM or query languages. It bridges the gap between human intent and machine-readable queries, democratizing access to BIM data for decision-making and analysis. By doing so, it demonstrates the power of combining semantic graph structures with LLMs to support intelligent, intuitive, and collaborative building analysis.

Since this research project was comparably small, the future research suggested by Massafra et al. (2024) includes the extension to bigger scales like digital twins.

Construction Waste Quantification

Construction waste (CW) quantification and optimization are essential for sustainable project delivery. However, inconsistencies in terminology, lack of standardized measurement units, and fragmented data hinder seamless automation and integration into early design and decision-making processes.

Sivashanmugam et al. (2024) address these limitations by proposing a BIM-integrated semantic framework that combines Building Information Modelling (BIM) with Semantic Web Technologies (SWT) to unify data structures, enhance interoperability, and support CW quantification and optimization across the building lifecycle. To develop this framework, the authors employed the Simple Knowledge Engineering Methodology (SKEM) and used the software Protégé to construct domain-specific ontologies. Two key ontologies called the Building Materials Ontology (BMO) and the Materials Properties Ontology (MAT), were populated with material, waste, and environmental impact data.

The data was drawn from reputable sources, including Environmental Product Declarations (EPDs), the European Waste Catalogue (EWC), BRESmartWaste, and other Lifecycle Assessment (LCA) databases.

To ensure consistent data classification, the Uniclass 2015 system is adopted, providing a hierarchical and structured coding scheme for construction elements and materials (NBS, n.d.). The resulting ontologies capture detailed instances of commonly used structural materials such as bricks, timber, plasterboard, steel, and insulation. Each material entry includes information on its technical attributes like volume, area and type of material.

The second ontology created is called PRODOCIRO. Its structure includes hierarchically represented waste data like Global Warming Potential (GWP) and Waste Generation Factors extracted from respective databases. These two ontologies are mapped together using the Uniclass nomenclature (Figure 5).

To be able to use and interact with the semantic web created, the researchers developed a BIM-based interface called the Building Waste Tool (BWT) within Autodesk Revit®, using Dynamo, Python, and SPARQL scripts.

The tool allows users to evaluate multiple material options based on both waste volume and environmental impact, thereby supporting waste-informed decision-making at the earliest design stages. Associated carbon impacts are visualized in tabular and graphical formats directly within the BIM environment. This enables real-time feedback and design iteration based on sustainability criteria.

Ultimately, the framework bridges existing gaps between BIM and CW data by transforming static design models into semantically enriched, interoperable systems.

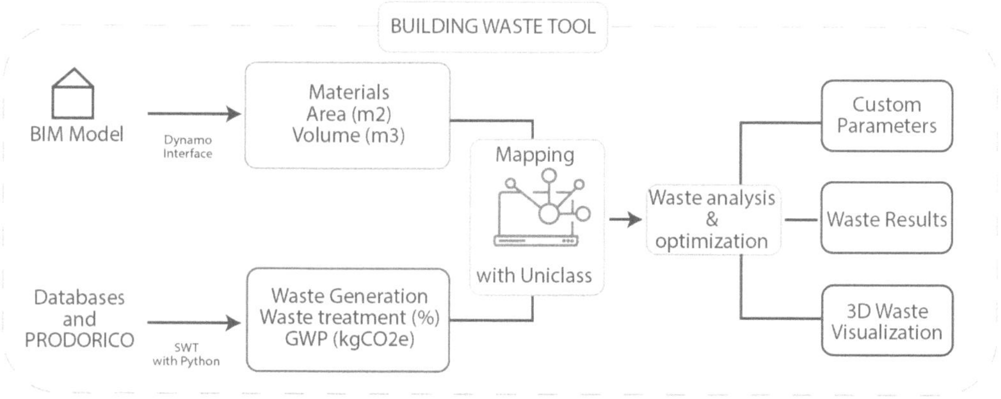

Figure 5
User Interface Workflow of BWT adapted from Sivashanmugam et al. (2024).

CONCLUSION

This paper has demonstrated that Semantic Web Technologies (SWTs) have the potential to fundamentally transform Building Information Modelling (BIM) in the AEC industry. By embedding semantic connections, SWTs address persistent challenges such as data fragmentation, limited interoperability, and inefficiencies in cross-disciplinary collaboration.

The three case studies presented, focusing on co-design workflows, natural language querying, and construction waste quantification highlight the versatility and impact on project efficiency of SWTs when integrated into BIM environments.

Despite promising outcomes, these approaches have primarily been tested on limited or small-scale datasets. Future research should therefore explore their scalability and robustness in real-world projects. Additionally, the integration of AI and Large Language Models (LLMs) offers a promising path to automate ontology and query generation with the goal to further bridge the gap between human users and machine-readable data. Advancing these areas will be key to unlocking the full potential of BIM as a semantically enriched, intelligent framework for sustainable and collaborative project delivery.

REFERENCES

BuildingSMART. (2024, November 15). *Industry Foundation Classes (IFC) - BuildingSMART International.* buildingSMART International. Retrieved May 26, 2025, from https://www.buildingsmart.org/standards/bsi-standards/industry-foundation-classes/

Elshani, D., Lombardi, A., Hernandez, D., Staab, S., Fisher, A., & Wortmann, T. (2025). *AEC Co-design workflow for cross-domain querying and reasoning using Semantic Web Technologies*. Automation in Construction, 176, 106226. https://doi.org/10.1016/j.autcon.2025.106226

Massafra, A. C. (2024). *Building Information Model Analysis Through Large Language Models and Knowledge Graphs.* In Data-Driven Intelligence - Proceedings of the 42nd eCAADe Conference, Nicosia, 685–694. https://papers.cumincad.org/cgi-bin/works/paper/ecaade2024_297

NBS. (n.d.). *Uniclass, delivered by NBS*. Retrieved June 10, 2025, from https://uniclass.thenbs.com/

Pauwels, P., Zhang, S., & Lee, Y.-C. (2017). *Semantic web technologies in AEC industry: A literature overview.* Automation in Construction, 73, 145–165. https://doi.org/10.1016/j.autcon.2016.10.003

Sivashanmugam, S., Rodriguez Trejo, S., & Rahimian, F. (2024). *BIM-integrated semantic framework for construction waste quantification and optimisation*. Automation in Construction, 168, 105842. https://doi.org/10.1016/j.autcon.2024.105842

W3C. (2014). *RDF 1.1 Concepts and Abstract Syntax.* World Wide Web Consortium. https://www.w3.org/TR/rdf11-concepts/

Optimizing Lighting in Work Environments

Simulation and optimization techniques

Amélie Assmayr[1]
Technical University Vienna
e12007770@student.tuwien.ac.at

This article talks about methods to simulate and optimize lightning which is an important topic in architecture and planning. The lighting standards an requirements will be presented followed by the presentation of three techniques to simulate or optimize lightning, namely the Monte Carlo method, Genetic algorithms and the response surface method. Lastly, the tools Radiance and DiaLux, which can be used to simulate lightning, will be shortly explained.

Keywords: *Lighting Simulation, Optimization, Monte Carlo, Genetic Algorithm*

INTRODUCTION

Lightning plays an important role in architectural design and planning. In workplace settings proper lightning can influence the comfort, wellbeing, and productivity of workers in offices. On top of that efficient lightning use can also be more energy efficient, which is why it is crucial to take it into account when designing workplaces. There are different tools and methods that can help designers to model and analyze lighting scenarios. This paper will explore the application of simulation and optimization techniques to improve lighting conditions in work environments (Gendler and Shojaee Barjoe, 2024)

LUX

Lux measures the number of lumen, which is the perceived power of visible light emitted by a source, per square meter (lm/m^2). 1 lux could, for example, be measured from a candle at half a meter distance. 100,000 lux outside on a very sunny day (Lampe Shop). The National Metrology Institute of Germany can realize illuminance levels from 0.001 lx to 100,000 lx. "A flux of 1000 lumens, spread uniformly over an area of 1 square metre, lights up that square metre with an illuminance of 1000 lux. However, the same 1000 lumens spread out over 10 square metres produces a dimmer illuminance of only 100 lux." (Wikipedia)

LIGHTING STANDARDS AND REQUIREMENTS

Since 2003, there have been standard requirements for lighting planning that must be fulfilled in Europe for indoor workplaces. In 2021, new rules were formulated and have been in effect since 2022. In European workplaces, lighting standards are primarily ruled by the guideline EN 12464-1, which contains minimum requirements for indoor workplace lighting including requirements for illuminance levels, glare control, and color rendering (Amrax). Office workstations should maintain an illuminance of at least 500 lux. Ceiling lighting and vertical lighting on the wall should have at least 30–150 lux. Higher values would be even better. The cylindrical illuminance, which is the light around people's faces, should be between 50–150 lux.

SIMULATION AND OPTIMIZATION METHODS IN LIGHTING DESIGN

There are a number of computational methods that can be applied in architectural lighting design. These include Monte Carlo simulation, genetic algorithms, and Response Surface Methodology, among others. In the following, I will describe the three methods and how they can be used in lighting simulation.

Monte Carlo method

Monte Carlo simulation is a stochastic technique used to solve physical and mathematical problems by simulating random variables. They are a good alternative to machine learning models when there is not enough data and are very accurate, which makes them helpful for long-term predictions. In lighting design, Monte Carlo simulations can be used for ray-tracing algorithms.

In lighting simulation, Monte Carlo ray tracing involves casting a large number of random rays from a light source and tracking their interactions with surfaces, including reflection, refraction, and diffusion. These randomly sampled rays simulate how light might reflect off other surfaces and bounce into the point being viewed.

For the simulation, you first have to determine "how many rays to send out in the Monte Carlo sampling of indirect lighting, how closely to space these values, how densely to sample the image plane, and so on." (J. Ward, 1994)

J. Ward used backward ray tracing. This means that each ray starts from a point on a surface (or a virtual sensor) and is traced backwards toward light sources. When a ray hits the surface, Monte Carlo sampling is used to estimate how much indirect light reaches that point. Monte Carlo simulations are helpful because they provide an efficient and flexible way to approximate the effects by averaging the results of many sampled rays.

An advantage of the Monte Carlo method is its high accuracy and flexibility. It can simulate a wide variety of material properties. On the other hand, Monte Carlo methods require a large number of samples to converge on an accurate result. This makes simulations slow without acceleration techniques like quasi-random sampling, path tracing, or photon mapping. A second disadvantage is that it is based on random sampling, which means that Monte Carlo simulation introduces variance in the output (noise), especially when using a low number of samples.

A tool that uses the Monte Carlo method for lighting simulation is, for example, Radiance, which was created by J. Ward. In the next chapter, this tool will be presented shortly..

Genetic Algorithms

A genetic algorithm is an optimization algorithm that is based on the process of natural selection in biology. They are used to solve complex optimization and search problems by mimicking evolutionary principles like mutation, crossover (recombination), and selection. Their advantage is that they are easy to understand, their computation can be parallelized, and they can solve problems that would be too complex for other algorithms. They have a lot of different applications and can also be used for lighting simulation (AnalytixLabs, 2024). "To implement a GA, three main subjects are required: definition of the chromosome structure, definition of a fitness function to evaluate possible solutions and to choose strategic genetic operators." (Mendes de Lima et al., 2010) The disadvantage is that this fitness function has to be carefully drafted to lead to good results. It analyzes how good each chromosome is as a possible solution.

In the paper by Cardoso and Lamounier (2010), each solution to the lighting layout problem is encoded as a chromosome composed of multiple genes, where each gene represents a specific attribute of the lighting system. In total, each chromosome encodes a full lighting configuration across several variables. The fitness function evaluated each configuration based on four criteria: shadow area, over-illumination, energy consumption, and overall system cost.

Response Surface Methodology

An effective method for optimizing workplace lighting conditions - especially in complex, multi-variable environments - is the integration of lighting simulation with Response Surface Methodology (RSM). RSM is a statistical and mathematical technique used for modeling and analyzing problems in which several independent variables influence one dependent variable. The objective is to optimize this response. RSM builds an approximate model (typically a polynomial) to represent the relationship between input variables and one or more responses. In their 2024 study, Barjoee and Gendler (2024) combined DIALux (see chapter Simulation Tools) lighting simulation with a Central Composite Design (CCD) statistical approach to optimize artificial lighting in a ceramic production facility. "CCD is a popular experimental design (DOE) used in response surface methodology (RSM) to optimize illumination" (Gendler and Shojaee Barjoee, 2024). It combines factorial points with axis and center points to allow analysis of the factor space. The total number of runs is given by Number of Runs = 2^k + 2k + n_c, where k is the number of factors and n_c the number of center points.

Using DIALux, the researchers created detailed 3D simulations of 11 distinct workspaces and calculated average, minimum, and maximum illuminance values. These simulated results were then validated through real-world lux measurements, with strong agreement between the two ($R^2 \approx 0.99$). For the optimization phase, the authors used Design-Expert software to explore the influence of four key lighting parameters—luminaire height, light loss factor, number of luminaires, and lighting technology (fluorescent vs. LED)—on visual performance. The method enabled the identification of the most impactful variable (lighting technology) and produced configurations with high desirability scores (>0.9), indicating optimal lighting uniformity and compliance with workplace standards. This study illustrates how simulation-based design, combined with statistical modeling techniques like RSM, offers a powerful framework for enhancing visual comfort and energy performance in real-world work environments (Barjoee and Gendler, 2024).

There are many more simulation and optimization techniques that can be used for lighting simulation and optimization. Depending on the goal and the data, other techniques might be more helpful, but this must be decided individually. These three techniques are general techniques that are also known from other areas, which is why they might be helpful for people who are not familiar with the topic yet.

SIMULATION TOOLS

In the chapter before, we talked about different techniques for lighting simulation. To do simulations, it is probably the easiest to use a preexisting tool. In this chapter, I will describe two tools for lighting simulations that are well known and often used. Of course, there are several more tools that might also employ other techniques, but this would go beyond the scope of this short paper.

Radiance

The Radiance lighting simulation system, developed by Greg Ward (1994), is the most widely used tool employing Monte Carlo ray tracing in architecture and daylighting. Radiance is capable of highly accurate simulations because it calculates both direct and indirect lighting using a backward Monte Carlo method. Radiance has served as the core engine for many other daylighting tools (Ward, 1994).

DiaLux

DIALux evo is a widely used free lighting design software—especially popular among architects and engineers—for both artificial and daylight planning. Inspired by Radiance but more accessible, it combines features such as photometric precision, daylight and electric light integration, automatic compliance checks with EN 12464-1, IESNA, and other regional norms, good 3D visualization, and many other features (DIALux).

CONCLUSION

Optimal lighting is an important topic, as incorrect lighting can lead to discomfort or inefficient energy use. Monte Carlo ray tracing can result in accurate assessments of lighting conditions. Optimization techniques like Genetic Algorithms or Response Surface Methodology can also be used in the design process by automating the search for ideal solutions. There are several preexisting tools that can make one's life easier and that employ some of these methods. Together, these tools ensure that workplace lighting meets not just regulatory standards but also promotes human wellbeing and productivity and is energy efficient. This paper just showed some examples, but there are many more tools and methods that can be used. As lighting design continues evolving, future directions will likely integrate real-time sensing, adaptive systems, and AI-driven optimization. This makes understanding and applying simulation and algorithmic tools more essential than ever.

References

Shojaee Barjoee S, Gendler S. Sustainable illumination: Experimental and simulation analysis of illumination for workers wellbeing in the workplace. Heliyon. 2024 Nov 30;10(24):e40745. doi: 10.1016/j.heliyon.2024.e40745. PMID: 39720084; PMCID: PMC11665526.

F. Mendes de Lima, I. Peretta, A. Cardoso and E. Lamounier, "Optimization of lighting design using Genetic algorithms," 2010 9th IEEE/IAS International Conference on Industry Applications - INDUSCON 2010, Sao Paulo, Brazil, 2010, pp. 1-6, doi: 10.1109/INDUSCON.2010.5740021

Bahdad, Ali & Syed Fadzil, Sharifah & Taib, Nooriati. (2020). Optimization of Daylight Performance Based on Controllable Light-shelf Parameters using Genetic Algorithms in the Tropical Climate of Malaysia. Journal of Daylighting. 7. 122-136. 10.15627/jd.2020.10.

Christoph F. Reinhart, Oliver Walkenhorst, Validation of dynamic RADIANCE-based daylight simulations for a test office with external blinds, Energy and Buildings, Volume 33, Issue 7, 2001, Pages 683-697, ISSN 0378-7788, https://doi.org/10.1016/S0378-7788(01)00058-5.

Gregory J. Ward. 1994. The RADIANCE lighting simulation and rendering system. In Proceedings of the 21st annual conference on Computer graphics and interactive techniques (SIGGRAPH '94). Association for Computing Machinery, New York, NY, USA, 459–472. https://doi.org/10.1145/192161.192286

Barjoee, S. S., & Gendler, S. (2024). Sustainable illumination: Experimental and simulation analysis of illumination for workers' wellbeing in the workplace. *Heliyon, 10*, e40745. https://doi.org/10.1016/j.heliyon.2024.e40745

Sivakumar, Ajithkannan. (2024). Mastering DIALux for Lighting Design and Calculation Author.

Wikipedia contributors. (n.d.). *Lux*. Wikipedia. Retrieved June 10, 2025, from https://en.wikipedia.org/wiki/Lux

Lampe Shop. (n.d.). *Wofür steht Lux?* Lampe Shop Magazin. Retrieved June 10, 2025, from https://www.lampe-shop.at/magazin/wofuer-steht-lux/?srsltid=AfmBOooTV_N8GMB2FdDQryyzpO7R9UKGHjFS3fcLmHmNCn3YA7W0MPXU

Amrax. (n.d.). *European indoor workplace lighting standards*. Amrax. Retrieved June 10, 2025, from https://amrax.ai/blog/european-indoor-workplace-lighting-standards/

DiaLux (n.d.) https://www.dialux.com/en-GB/dialux. Retrieved June 10, 2025

Chen, J., An, J., Yan, D. *et al.* Future trends in intelligent lighting control systems: Integrated technologies, multi-system linkage and AI integration. *Build. Simul.* **17**, 1909–1932 (2024). https://doi.org/10.1007/s12273-024-1209-3

Techtarget, https://www.techtarget.com/searchcloudcomputing/definition/Monte-Carlo-simulation Retrieved June 10, 2025

AlayticsLabs, Medium, https://www.analytixlabs.co.in/blog/genetic-algorithm/#Running_a_Genetic_Algorithm_in_Python Retrieved June 10, 2025

NeuralConcept, https://www.neuralconcept.com/post/response-surface-methodology-overview-applications#link1 Retrieved June 10, 2025

Space Gene and Generative Urban Design

A theoretical framework to research Urban Morphology

Yang Yucheng
TU Wien
e12402335@student.tuwien.ac.at

The theory of space gene, proposed by Duan Jin, conceptualizes urban morphology as the result of long-term interactions between natural, cultural, and historical forces, encapsulated into a stable and locally distinctive spatial structure. With the rise of algorithmic planning and data-driven urban analysis, this theory now finds renewed relevance. This review critically synthesizes the evolution of spatial gene theory within the context of quantitative urban morphology and explores its integration into generative urban design practices. By synthesizing theory, method, and design, this paper positions spatial gene not only as a cultural construct but also as an operational planning tool. It calls for future work that connects deep morphological structures with adaptive and localized algorithmic planning models, to address challenges of identity loss and spatial homogenization in rapidly urbanizing contexts.

Keywords: *Quantitative Urban Morphology, Space Gene, Generative Urban Design, Algorithmic planning, Urban fabric classification*

INTRODUCTION

Spatial genes represent the fundamental interplay among the morphological characteristics of historical districts. Based on the research of morphological types, in 2019, academician Duan Jin proposed the "spatial gene" theory for use in urban design. He concluded that space gene is a unique and relatively stable space model formed by the interaction between urban space, natural environment, and history and culture. It conveys distinctive local characteristics of different regions, form a special city identity, and maintains the harmonious relationship between urban space, natural environment, and history and culture.

Currently, accelerated urbanization and prosperous economic growth stimulate the urbanization process swiftly and significantly. However, it also leads to the homogenization of urban landscape. The identification, analysis, and guidance techniques of space gene can improve the localization of urban planning and design and finally mitigate the problem of homogenization.

THE EMERGENCE OF QUANTITATIVE MORPHOLOGICAL ANALYSIS METHODS

With the advancement of information technology, data-driven urban management approaches have continuously emerged, not only optimizing resource allocation but also identifying and responding to the needs of various population groups in a timely manner. These approaches provide a data foundation for improving urban spatial quality and enhance the efficiency of urban governance. Compared to traditional urban planning, big data-based urban studies are "quantifiable, measurable, and monitorable," transforming static spatial planning into dynamic spatiotemporal planning and making urban governance more targeted.

Urban morphology typology plays a critical role in understanding the spatial structure of cities. However, existing research often relies on qualitative classifications or single clustering algorithms, which are subject to strong subjectivity and questionable reliability (Li et al, 2024)

Driven by the rise of new urban science and the increasing need for precision in urban research, quantitative morphology has emerged alongside advances in computing technologies and the availability of multi-source data. In recent years, the development of big data and new technologies has opened new opportunities for quantitative urban morphology research. The emergence of multi-source data and analytical methods has enabled fine-scale, human-centred studies, offering new approaches for objective measurement of urban form and the analysis of its effects (Gil et al., 2011).

For example, deep learning models trained on large urban fabric datasets are now used for the automated classification of urban spaces, helping to uncover the spatiotemporal evolution of urban structures (Zhou et al., 2024). The emergence of such methods provides increasing technical support for quantitative morphological analysis.

THE DEVELOPMENT OF SPACE GENE THEORY

The term "spatial gene" refers to the unique information encoded in urban spaces that emerges from long-term interactions among the natural environment, historical context, and urban development. It serves both as an identifier of local urban character and as a means to maintain the harmony between these three aspects (Fleischmann et al., 2021).

Although the concept of Space Gene was firstly proposed by Duan Jin in 2019, it derives from several theories. The idea of applying genetic concepts to urban morphology originated in 19th-century Germany, focusing on the formation and evolution of urban form. The classification of urban patterns based on geometric similarity is conceptually similar to the morphological taxonomy used in early biology. While modern biological taxonomy relies on DNA sequencing and phylogenetics, morphological analysis still provides valuable insights for studying urban form.

Just as phenotypic and morphological descriptions help distinguish biological species into homogeneous groups, numerical taxonomy can be extended to urban form to offer a feasible and reliable conceptual and methodological framework for the systematic classification of homogeneous urban morphologies.

Prior to the introduction of the "spatial gene" theory, concepts such as "cultural gene" and "landscape gene" focused primarily on preserving historical cultural features, with limited attention to spatial analysis. The theory includes techniques for identifying, evaluating, and guiding the inheritance of spatial genes, addressing the design failures that arise from applying uniform models to diverse urban contexts without respect for history or nature. This approach enhances contextual sensitivity in planning and facilitates a shift from form-making to spatial gene analysis, offering a practical design strategy for harmonizing urban development with natural preservation and cultural continuity.

URBAN FABRIC IDENTIFICATION BY SPACE GENE APPROACHES

Urban fabric is a complex system characterized by its connectivity, interaction rules, and hierarchical organization. Identifying urban fabric is foundational to its classification. However, challenges such as data quality, consistency, and availability, as well as ambiguity in defining urban morphology and its constituent elements, continue to hinder the development of quantitative analysis methods. Therefore, clearly defining the basic elements of morphological research and ensuring their measurability are prerequisites for quantitative studies.

Various data sources are used for fabric identification, including maps and remote sensing imagery. GIS-based methods often utilize OpenStreetMap (Zhou et al., 2024) and European urban map datasets (Mascherek et al., 2024) to acquire information on buildings and street networks. Remote sensing provides high-resolution images of urban areas for extracting features like buildings and vegetation to study spatial structure and form (Arribas-Bel et al., 2022) (Boeing, 2019).

The unit of analysis also affects identification results. The classical triad in morphology—streets, plots, and buildings—are often used as basic analytical units. Plot-based classification is a common method in urban morphology research. For instance, Na Li et al. (2024) developed a knowledge-based classification framework combining planning knowledge and machine learning to identify residential block types in Seoul.

Building-based classification approaches have also emerged, using data such as location, shape, and area. These approaches analyze relationships between individual buildings within urban networks (Wang et al., 2023). Other novel units include the "enclosed tessellation unit" (ETU), which meets the criteria of indivisibility, internal consistency, and spatial exhaustiveness. It captures both the physical characteristics of bounded landscapes and the spatial relationships between buildings, avoiding the rigidity of administrative or uniform grid-based divisions (Arribas-Bel et al., 2022).

Currently, classification methods for urban fabric primarily focus on six categories: dimension, shape, spatial distribution, development intensity, connectivity, and diversity.

While significant progress has been made in urban fabric research, challenges remain in defining basic elements and ensuring their quantifiability. Further methodological refinement is needed to enable more effective analysis of urban fabric.

URBAN FABRIC CLASSIFACATION BY SPACE GENE APPROACHES

The goal of classifying urban fabric is to reduce complexity and gain a better understanding of urban organizational patterns. Beginning with simple binary classifications, methods have evolved into more sophisticated systems. The development of complex network science offers new perspectives for quantitatively analyzing urban fabric through spatial networks. Traditional classification approaches—such as expert-scoring-based assessments—can incorporate historical, cultural, and social dimensions for comprehensive analysis (Li et al., 2024). However, such methods are limited by the subjectivity of expert knowledge and experience.

Emerging deep learning-based classification methods, including Convolutional Neural Networks (CNNs), are capable of learning urban fabric features and extracting representative feature vectors.

Network-based classification methods have also become prominent (Fleischmann et al., 2022), including K-means clustering, hierarchical clustering, and Bayesian network clustering (Wang et al., 2023). These are well-suited for data with hierarchical structures—such as buildings, parcels, and street networks—and can visually represent hierarchical relationships and clustering through tree structures.

By examining the evolution of classification methods, we observe a shift from simplicity to complexity and recognize the potential of network-based clustering methods—particularly hierarchical clustering—as effective tools for understanding urban fabric as a complex system. These approaches help reduce complexity and provide insights into urban organization patterns.

GENERATING URBAN DESIGN IN PRACTICE

Recently, several generating urban design studies based on the space gene concept have already been implied globally in different aspects, which indicates the huge potential of the theory. Duarte & Beirão proposed a flexible urban design methodology using combination of Urban Patterns and Shape Grammars (Duarte & Beirão, 2011). They offer a robust framework for generating adaptable urban forms based on predefined rules and contextual patterns, aligning with the goals of flexibility and responsiveness in contemporary planning practice. In the quarter scale, Koenig et al. (2019) present a modular optimization framework for urban design using evolutionary algorithms. The case study in Weimar, Germany, illustrates the approach, producing a set of Pareto-optimal solutions. In the block scale, Makki et al. (2015) explored the use of evolutionary algorithms in urban design in Barcelona Eixample. They stimulated how evolutionary design strategies can reinterpret Cerda's original urban ideals and underlined the adaptability and flexibility of multi-objective evolutionary computation in architecture.

Except for the urban design, the generating urban design also plays an important role in Microclimate optimization. Xu et al. (2019) proposed a computer-aided, genetic algorithm-driven optimization method for urban block design, focusing on improving outdoor thermal comfort in dry and hot climates. In Kashgar, China, the authors simulate and optimize urban layouts based on Universal Thermal Climate Index (UTCI). This research offers design guidance and a simulation-optimization framework to improve urban climate adaptability in arid regions.

FROM SPACE GENES TO PLANNING IMPLEMENTATION: POLICY-DRIVEN APPLICATIONS

To bridge the theoretical framework of spatial genes with concrete planning interventions, recent studies and pilot projects have begun to explore how the classification and extraction of spatial genes can support regulatory planning, heritage conservation, and zoning policy formulation. This section outlines three practical strategies that demonstrate how spatial gene theory can transition from analytical insight to operative planning support.

Spatial Gene-Informed Regulatory Planning in Historic Urban Areas has already been implied in China (Tao & Ye, 2022). In Nanjing's Old South District, a pilot project led by Southeast University applied space gene extraction techniques to inform urban renewal guidelines. By identifying dominant morphological patterns—such as block dimensions, building setbacks, and street hierarchies—using clustering and remote sensing data, planners developed zone-specific design regulations that respected historical continuity while enabling infill development. These regulations included: Plot-level density controls based on identified genetic typologies; Street façade continuity rules tied to historical enclosure patterns; Height gradient zones respecting the skyline heritage of adjacent monuments. The morphological features were extracted using remote sensing imagery and unsupervised clustering techniques, enabling planners to map spatial typologies with high granularity. The project employed a multi-source GIS framework to integrate historical maps, land-use layers, and built-form indicators, thereby enhancing the reliability of zoning recommendations. This approach demonstrates how spatial gene theory can be translated into control-based planning instruments, enhancing cultural sensitivity and spatial coherence in densifying urban cores.

In Matera, Italy, a geometric layer as a foundational component of Urban Digital Twins (UDTs) is proposed, emphasizing how 3D morphological representations of cities can support analytics, simulations, and urban policy tools by decoding spatial patterns, which can be interpreted as morphological traits. The digital twin model was constructed using LiDAR-based 3D mesh generation, enriched by semantic annotation of urban elements (e.g., building function, height, and typology) for simulation and analysis (De Giglio et al., 2023) . This enabled real-time environmental assessments and morphological scenario modeling, laying a foundation for regulation-sensitive urban transformation. The traits can guide simulations and regulatory compliance. This article supports the idea that urban form can be classified, embedded, and governed through geometric layers—making it an ideal technical companion to spatial gene theory. It highlights how form-based urban data, when structured and semantically linked, can inform real-time urban decisions and long-term morphological conservation (Scalas et al., 2022).

DISCUSSION AND CONCLUSION

This review has demonstrated that spatial gene theory, when integrated with quantitative morphology and algorithmic design methods, holds significant potential for addressing the dual challenge of spatial homogenization and identity loss in contemporary urbanization. By grounding urban form classification in both cultural memory and computational logic, spatial genes can serve as the foundation for adaptive, localized, and context-sensitive urban design strategies.

The spatial gene theory offers a practical framework for translating urban morphological patterns into planning tools. Recent applications demonstrate how spatial gene extraction can inform regulatory design, preserve local identity, and support adaptive planning through digital platforms.

These cases show that spatial genes are not merely theoretical constructs, but scalable instruments for context-sensitive planning. To maximize impact, future research should focus on standardizing data inputs, improving classification interoperability, and embedding spatial equity into morphological governance—thus reinforcing the role of spatial genes in algorithmic, identity-aware urban planning.

REFERENCES

Arribas-Bel, D., & Fleischmann, M. (2022). Spatial Signatures – Understanding (urban) spaces through form and function. *Habitat International, 128*, 102641. https://doi.org/10.1016/j.habitatint.2022.102641

Boeing, G. (2021). Spatial information and the legibility of urban form: Big data in urban morphology. *International Journal of Information Management, 56*, 102013. https://doi.org/10.1016/j.ijinfomgt.2019.09.009

De Giglio, M., Spano, M., Bruno, G., & Valvano, V. (2023). Digital twins for intelligent cities: The case study of Matera. *Journal of Reliable Intelligent Environments*. https://doi.org/10.1007/s40860-025-00245-3

Duarte, J. P., & Beirão, J. N. (2011). Towards a methodology for flexible urban design: Designing with urban patterns and shape grammars. *Environment and Planning B: Planning and Design, 38*(5), 879–902. https://doi.org/10.1068/b37026

Fang, Z., Jin, Y., Zheng, S., Zhao, L., & Yang, T. (2024). UrbanClassifier: A deep learning-based model for automated typology and temporal analysis of urban fabric across multiple spatial scales and viewpoints. *Computers, Environment and Urban Systems, 111*, 102132. https://doi.org/10.1016/j.compenvurbsys.2024.102132

Fleischmann, M., Romice, O., & Porta, S. (2021). Measuring urban form: Overcoming terminological inconsistencies for a quantitative and comprehensive morphologic analysis of cities. *Environment and Planning B: Urban Analytics and City Science, 48*(8), 2133–2150. https://doi.org/10.1177/2399808320910444

Fleischmann, M., Feliciotti, A., Romice, O., & Porta, S. (2022). Methodological foundation of a numerical taxonomy of urban form. *Environment and Planning B: Urban Analytics and City Science, 49*(4), 1283–1299. https://doi.org/10.1177/23998083211059835

Gil, J., Beirão, J. N., Montenegro, N., & Duarte, J. P. (2011). On the discovery of urban typologies: Data mining the many dimensions of urban form. *Urban Morphology, 16*(1), 27–40. https://doi.org/10.51347/jum.v16i1.3966

König, R., Miao, Y., Aichinger, A., Knecht, K., & Konieva, K. (2019). Integrating urban an alysis,generative design, and evolutionary optimization for solving urban design problems. *Environment and Planning B: Urban Analytics and City Science, 47*, 997-1013. https://doi.org/10.1177/2399808319894986

Li, J., & Li, C. (2024). Characterizing urban spatial structure through built form typologies: A new framework using clustering ensembles. *Land Use Policy, 141*, 107166. https://doi.org/10.1016/j.landusepol.2024.107166

Li, N., & Quan, S. J. (2024). Discovering urban block typologies in Seoul: Combining planning knowledge and unsupervised machine learning. *Cities, 150*, 104988. https://doi.org/10.1016/j.cities.2024.104988

Makki, M., Farzaneh, A., & Navarro, D. (2015). The evolutionary adaptation of urban tissues through computational analysis. In *Proceedings of the 33rd International Conference on Education and research in Computer Aided Architectural Design in Europe (eCAADe)*, Volume 2 (pp. 563–571). https://doi.org/10.52842/conf.ecaade.2015.2.563

Mascherek, A., Düzel, S., Eibich, P., Krekel, C., Goebel, J., Gallinat, J., Wagner, G. G., Lindenberger, U., & Kühn, S. (2024). Working memory and fluid intelligence are differentially related to categories of urban fabric in older adults: Results from the Berlin aging study. *Journal of Environmental Psychology, 93*, 102224.

Scalas, A., Cabiddu, D., Mortara, M., & Spagnuolo, M. (2022). Potential of the geometric layer in urban digital twins. ISPRS International Journal of Geo-Information, 11(6), 343. https://doi.org/10.3390/ijgi11060343

Tao, Y., & Ye, R. (2022). *Analysis of the Spatio-Temporal Characteristics of Nanjing's Urban Expansion and Its Driving Mechanisms. ISPRS International Journal of Geo-Information, 11*(7), 406. https://doi.org/10.3390/ijgi11070406

Wang, J., Fleischmann, M., Venerandi, A., Romice, O., Kuffer, M., & Porta, S. (2023). EO+Morphometrics: Understanding cities through urban morphology at large scale. *Landscape and Urban Planning*. Advance online publication. https://doi.org/10.1016/j.landurbplan.2023.104691

Xu, X., Yin, C., Wang, W., Xu, N., Hong, T., & Li, Q. (2019). Revealing urban morphology and outdoor comfort through genetic algorithm-driven urban block design in dry and hot regions of China. *Sustainability, 11*(13), 3683. https://doi.org/10.3390/su11133683

Zhou, F., Jin, Y., Zheng, S., Zhao, L., & Yang, T. (2024). UrbanClassifier: A deep learning-based model for automated typology and temporal analysis of urban fabric across multiple spatial scales and viewpoints. *Computers, Environment and Urban Systems, 111*, 102132. https://doi.org/10.1016/j.compenvurbsys.2024.102132

Floor Plan Object Recognition Using Machine Learning

New Application Strategies for Floor Plan Object Recognition Using Machine Learning: A Case Exploration of Foundation Models

Hayoung Lim
Vienna University of Technology
e12348271@student.tuwien.ac.at

Recent advancements in machine learning have accelerated the development of floor plan object recognition technologies.This paper reviews two widely used models—DeepFloorplan and a U-Net-based Floor Plan Parsing model—focusing on their core algorithms, features, and limitations. Based on this analysis, we explore how emerging tools such as the Segment Anything Model (SAM) and GPT-4 could address current challenges. The goal of this study is to assess the current state of floor plan recognition technology and suggest new directions for future development.

Keywords: *Floor Plan Recognition, Architectural Design Automation, Computer Vision in Architecture.*

Introduction

The development of architectural design automation has advanced quickly with artificial intelligence and digital design technologies. Specifically, automation tools are now being applied even more from the beginning of the design, changing the way we think about architectural design, creating efficiencies, minimizing errors, and allowing designers to make data-driven decisions.

This research focuses on floor plan object segmentation and recognition from the many processes of the realm of design automation. This technology intends to automatically recognize, segment, and label structural elements, like rooms, doors, windows, and furniture in a 2D architectural plan. The collaboration of segmentation and labels is the vital output of this early design task, as it drives downstream outputs, like space planning, code compliance, simulations, and accessibility evaluations.

As such, this work highlights the specialized limitations of existing approaches and presents a novel hybrid that uses advanced foundation models to help address these limitations. In doing so, the exploration and potential of these models will demonstrate another way to automate floor plan interpretation that is more flexible and scalable.

In this work DeepFloorplan and a U-Net based floor plan parsing model were chosen to represent the common approaches in this field. We will undertake detailed analysis of these models, highlighting their functioning and performance.

DeepFloorplan

DeepFloorplan is a model utilizing a multi-task neural network developed with the intention of being able to recognize and assist with the detection of major structural parts from 2D architectural floor plan images. The process of detecting the structural components ranges from

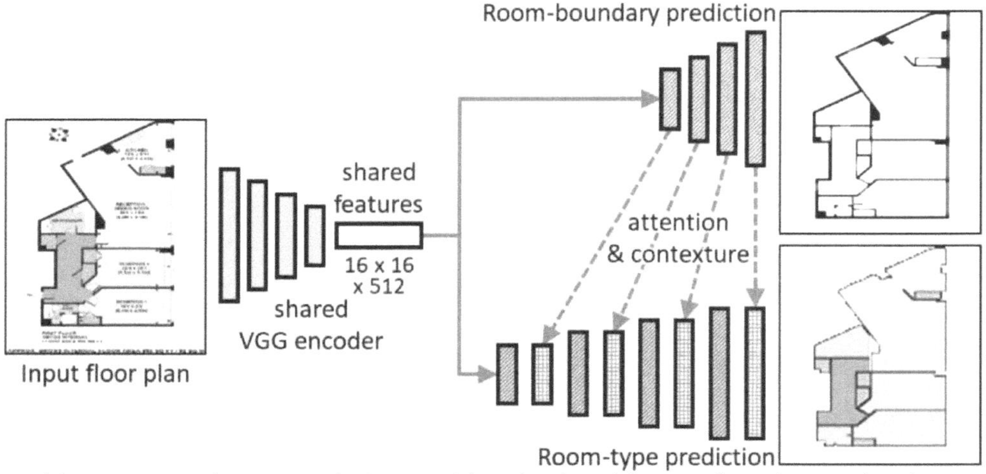

Figure 1 : Overall network Architecture of DeepFloorplan

just object segmentation to producing spatial configuration and functional layout at the same time while also helping the user understand the segmentation as well.

To begin, a single VGG encoder is used to obtain the same features of the input image with respect to the entirety of the structure. VGG is a deep convolutional neural network that was developed for the task of image classification. As it relates to DeepFloorplan, the VGG encoder will be used to extract visual representations of the 2D floor plan images. VGG is successfully utilized with very complex images such as floor plans, since the lines, pieces or segments are distinctly identified and they are good at finding sharp edges, structures, and spatial arrangements. Within the DeepFloorplan framework using VGG pretrained weights allows efficient transfer learning in modest training data situations.

After the encoder extracts data, it is processed and predicted by two decoders. The decoder used for the roof boundary is situated at the top and it predicts wall, door, and window boundaries at the pixel level based on the features provided by the encoder. The Room-Type Decoder below separates spatial functions—bedrooms, living rooms and kitchens, bathrooms—at the cell level while utilizing both the shared features from the encoder along with intermediate outputs from the top decoder.

The most distinctive architectural element of DeepFloorplan is the Room-Boundary-Guided Attention Mechanism. In this case it is guided attention where the Room-Boundary-Guided Attention Mechanism does not simply split the two decoders (as indicated in parallel), but instead uses the boundary information (walls, doors, windows) produced by the top decoder, and builds this information into the analysis of the bottom decoder that provides for structural contextual information in it predictions. The fundamental purpose is to think how each element - rooms, doors, walls, for example - are related not in isolation, but to the structure that surrounds them.

Nonetheless, despite this mechanism, DeepFloorplan does not seem to achieve a deep understanding of spatial semantics. Results from experimental testing suggest that while the model seems adept at predicting the common spatial patterns that are frequently modeled from the training data, it still struggles with more complex or atypical structures, and it does not recognize or misrecognizes components (c.f. constrained linear structures). Bent corridors, curved walls and new forms of room partitions

Figure 2 : Recognition error of DeepFloorplan

are often inaccurate in recognition. Further, non-structural features such as compass indicators or symbolic icons are usually misclassified either as part of walls or room categories either as walls or room types. This indicates that the DeepFloorplan model is based on visual similarity and repetition of data and does not rely on recognizing a contextualized spatial structure.

Furthermore, in DeepFloorplan the decoders need high-quality labeled data at the pixel-level. For example, to enable the Room-Boundary Decoder to determine pixel-level objects such as walls, doors, and windows, class information needs to be provided for each pixel, "this pixel is a window." In architectural drawings, a higher degree of labeled precision is required because even minor misalignment can yield a large error rate.

However, elements in floor plan images are difficult to delineate or be recognized accurately through automated means. Interpreting such drawings typically requires contextual knowledge of spatial functions and relations. This means that usually, manual boundary labeling of individual objects is carried out by human annotators with high precision. This takes considerable time and human resources. Constructing a large-scale database that accurately reflects the variety of floor plan structures and is labeled with high precision is certainly not an easy undertaking, especially as labeled precision has a high requirement. As a result, the non-diversity of data results in a corresponding decrease in the generalization performance of the model.

Consequently, a structural adjustment procedure through post-processing is necessary at the last step of the program. This requirement damages the efficacy of the end-to-end approach, bringing a significant obstacle to the goals of simplicity and real-time performance of automation technologies.

U-Net-based Floor Plan Parsing Model

U-Net is a segmentation network based on deep learning which was proposed in 2015. It is popularly used in image analysis problems which

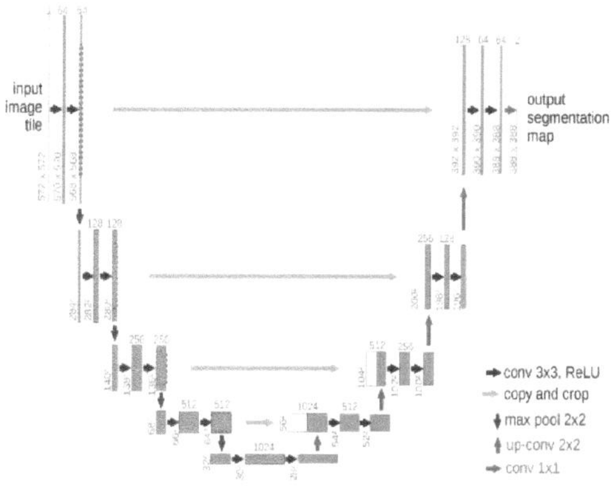

Figure 3 : Architecture of U-net

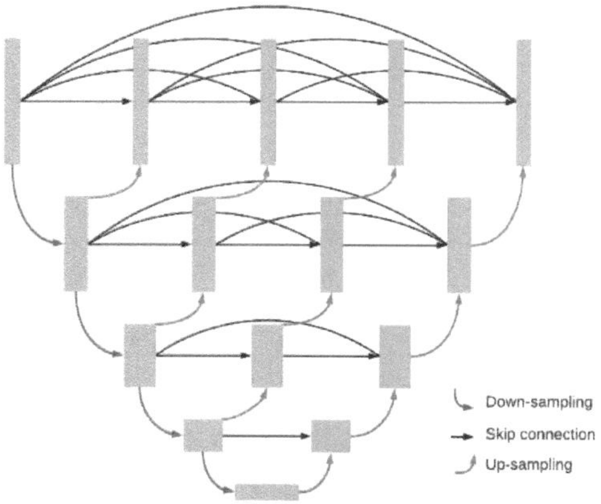

Figure 4 : Skip connection of U-net

require precise boundary detection such as architectural plan analysis and medical imaging. Although U-Net follows the general encoder-decoder structure, it is unique in that it utilizes skip connections to preserve high resolution pixel values which are required when needing to detect fine boundaries.

The general architecture of U-Net is U-shaped and symmetrical, with a contracting path on the left and an expanding path on the right. The two halves of the network are connected by the flow of information left to right. As the information moves from left to right the image is first compressed through the contracting path, then is then restored through the expanding path.

The contracting path uses 3×3 convolutional filters to extract features from a 2D floor plan image. ReLU is used as an activation function, which preserves important signals. After applying the first 3×3 filter, the resolution is reduced by 2×2 max pooling. The max pooling will help ensure that pixels represent more area when pooling, because it can only observe a smaller area from the original resolution of the image. Overall, the network's ability to learn spatial arrangement, structural context, and boundaries at only a local scale is compromised unless the resolution is somehow down sampled; in essence, the contracting path is a many-layered feature extraction technique that down-samples the image by condensing it in the sense that it is summarizing the most important information as feature maps.

However, these summarized features are still too abstract and do not maintain the pixel-level fidelity needed to characterize the built environment at the level of detailed spatial segmentation. Hence, up-sampling itself will not restore the separate and detailed representations within the respective class labels of the built environment. The expanding path takes advantage of both up sampling or transposed convolution to gradually restore and eventually bring an image back to its original size. Simply up-sampling- is impossible without losing the spatial positioning information. For this reason, U-Net is designed to consider the implications of spatial data loss, so that it can accommodate skip connections that expedite the delivery of high-resolution information from the contracting path to progressively later stages of the expanding path. There is a considerable difference between the condensed abstracted forms of spatial information (feature maps) at the end of the contracting path, and the high-resolution spatial, and boundary information for the image, before it is restored during the up-sampling process.

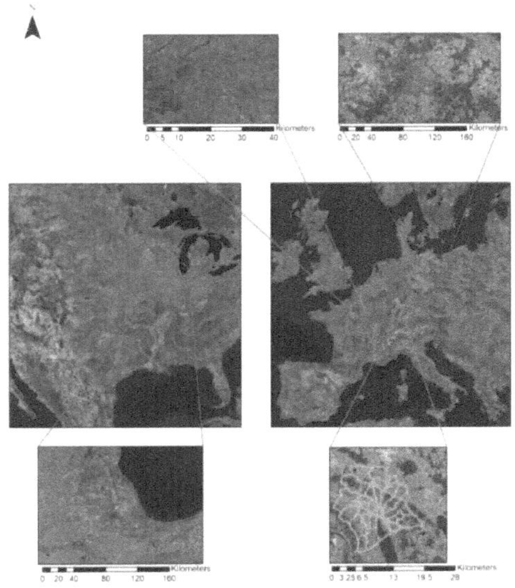

Figure 5:
High misclassification rate of U-net in densely built urban areas (Berlin)

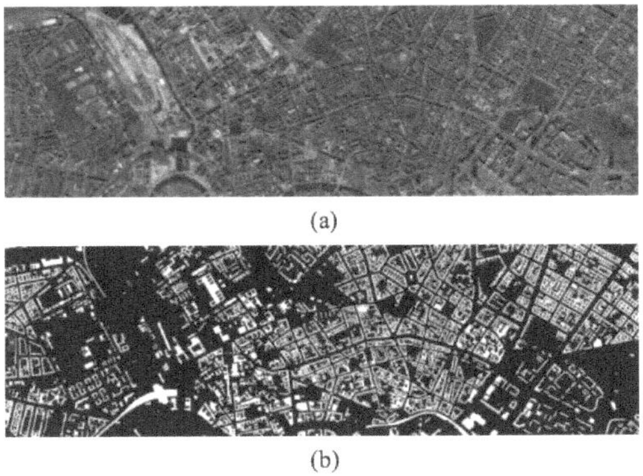

Figure 6:
High segmentation accuracy of U-net in sparsely built areas (Zurich)

However, U-net has a few architectural limitations. The first limitation is that the skip connections of the architecture work only as simple concatenations, which makes it much more difficult to resolve the information gap between the contracting and expanding paths. When we talk about an information gap, we are plainly discussing a semantic gap that results from the structural simplicity of the skip connections. The contracting path refers to the low-level information, like locations and boundaries, while the expanding path refers to higher-level semantic information, like shapes and meanings. When we can merge types of information into one step, we naturally may lose consistency and degradation of segmentation accuracy, especially with complex boundaries or poorly defined boundaries.

The second limitation is that U-net cannot natively handle multiple scales of an object well. The model must use one level of resolution for up-sampling at a time, which provides a major challenge to detecting shapes of different sizes and shapes in an image. For example, when the model is presented with multiple objects of different sizes and shapes, the model tends to home in on one object and ignore the rest. For this reason, segmenting every element within an image with the same degree of accuracy remains difficult. This limitation is accentuated in architectural floor plans that typically include different spatial structures, creating a considerable reduction in performance.

Thirdly, there are practical limitations regarding the availability of training data. While U-Net was specifically designed as a deep learning architecture that potentially performs effectively in a limited training data setting, achieving a high level of performance across very diverse domains in real-world practice still requires many thousands of high-resolution images, along with pixel-wise annotations (ground truth) that could be generated manually or through a crowdsourcing method. Unlike the medical field, where models are able to generalize well given even the limited training data, in practice architectural floor plans are highly varied and largely non-standardized in their structure.

The more complex and diverse spatial information and geometric forms in the training data will require an even larger number of examples before the model can generalize. Ultimately, this means these multi-level training examples, regardless of the cost, will require significant human effort (cost) and time to gather and annotate. However, as was done with DeepFloorplan, the annotations will be generated manually, and the cost is even higher in human effort (cost) and time.

This limitation represents a significant barrier to employing U-Net based segmentation systems in the practice of architecture in the real world.

GPT-integrated Multi-object Few-shot SAM (GMFS)

Both DeepFloorplan and U-Net are established deep learning-based models that have demonstrated superior performance on architectural floor plans (DeepFloorplan) and general semantic segmentation (U-Net) tasks. However, both share a fundamental architectural limitation in that they both require vast high-quality pixel-level-annotated datasets to be effective. As both models are attempting to understand complex spatial constructions and precise boundaries, the absence of consistent annotated data will lead to significant performance degradation.

DeepFloorplan requires precise pixel-level annotations to learn the complex configurations of rooms and many wall types. However, this type of annotation is manually drawn, thus there are restrictions on the availability of data. U-Net was initially designed for small-scale data environments, but it also requires thousands of high-resolution images and extremely detailed annotations to be able to generalize across many object shapes and irregular structures. This required precision will require a lot of time and financial cost, which emphasizes limitations once again. These extensive dependencies on data scale and quality are an important limitation for both models that significantly reduce their practical use.

Nonetheless, amidst fast-moving progress in deep learning, we are witnessing the emergence of pre-trained foundation models trained on large datasets that show potential for applicability instead of the expensive domain-specific models, which have limitations in terms of generalization and reliance on training data This chapter discusses our new approach to architectural floor plan analysis through the use of two representative foundation models: the Segment Anything Model (SAM) and ChatGPT. Then we discuss the potential benefits of this approach compared to approaches such as DeepFloorplan and U-Net.

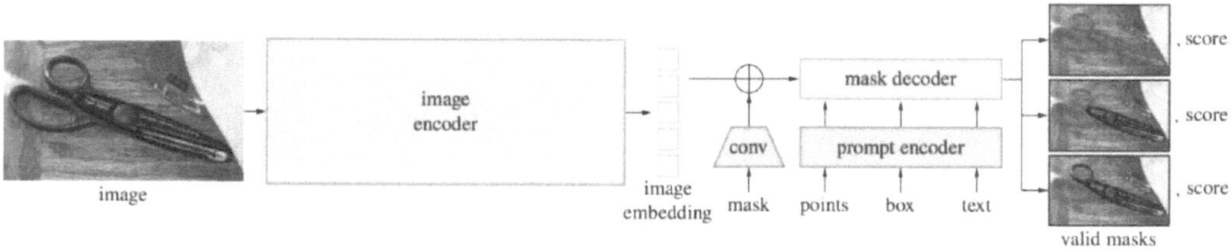

Figure 7 : Segmentation of SAM (Segment Anything Model)

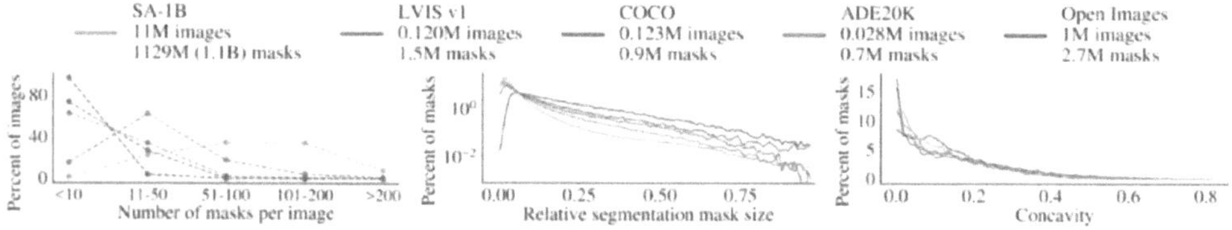

Figure 8 : Dataset mask properties of SAM (Segment Anything Model)

The Segment Anything Model is able to segment images based on a desired specifiable segmentation and construct empirically accurate areas. Furthermore, it can identify and segment various features and categories of objects in real-time with little to no user prompts. Users may indicate desired objects through simple point prompts, bounding box prompts, or text prompts - then SAM will output fast and relatively accurate segmentations. SAM was pretrained on a very large-scale dataset (SA-1B) of billions of image-mask pairs, allowing them to make meaningful comparisons in whatever domain, without retraining it. SAM shows remarkable flexibility in operating directly, even in visually complex and unstructured scenarios such as architectural floor plans which typically represent patches of space that are not only randomly but also imperfectly configured, with symbols, or with a diversity of objects with differing shapes and sizes.

ChatGPT, built on OpenAI's GPT-4 architecture, is a natural language processing model, yet is capable of human-level abilities to understand context and produce text that is contextually reasonable. Within the context of successful floor plan recognition, ChatGPT is capable of interpreting text, semantic labeling, and deducing relationships amongst structural elements, thus it can recognize text automatically when inserted into drawings and assign spatial functions accordingly. ChatGPT also provides names for structural elements or deduces high-level spatial structures with the visual data that has been produced from SAM. In addition, when a user's issues command in a natural language, ChatGPT understands, translates, and surprises various instructions in the system.

GMFS (GPT-integrated Multi-object Few-shot SAM) is a newly proposed hybrid system to reasonably interpret structured visual information that includes architectural floor plans. It integrates the Segment Anything Model (SAM) and OpenAI's GPT-4. Existing Deep Learning-based segmentation models require large amounts of manually labeled data, apply to segmentation of a single object, and lack mechanisms for semantic interpretation, whereas GMFS segments multiple objects and classifies

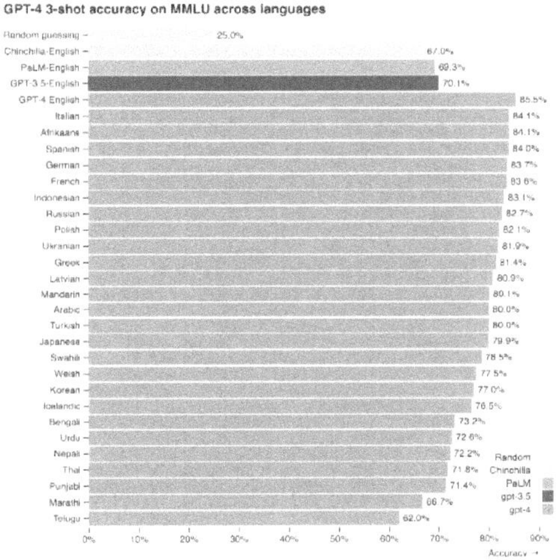

Figure 9 :
5. Performance of GPT-4 in a variety of languages

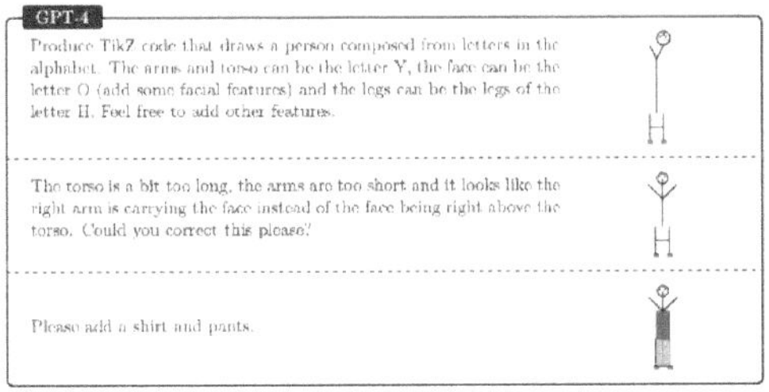

Figure 10 :
GPT-4's ability to incorporate user feedback

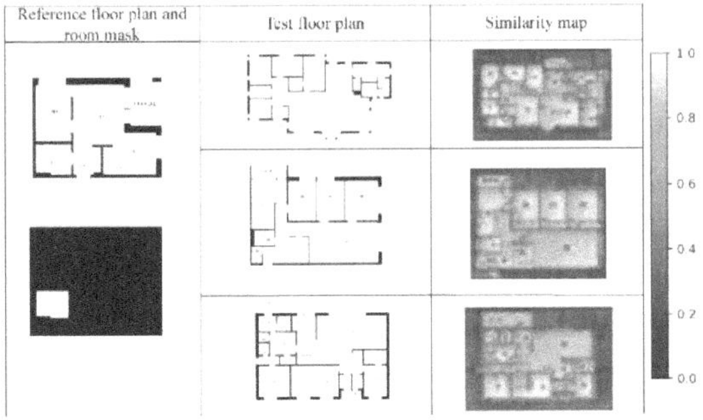

Figure 11 :
Similarity map

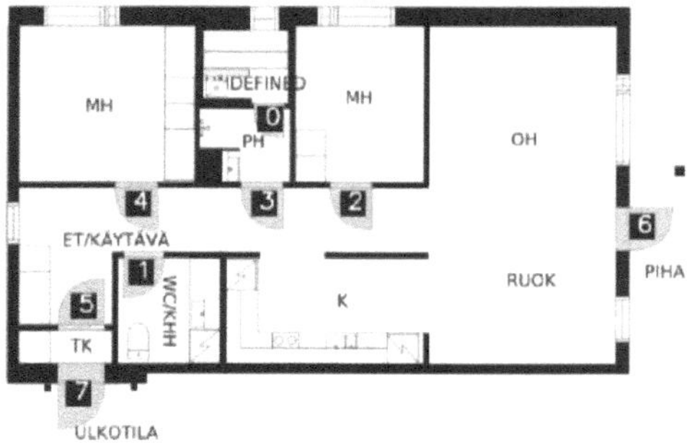

(a) Input Floor Plan with Indexed Door Masks

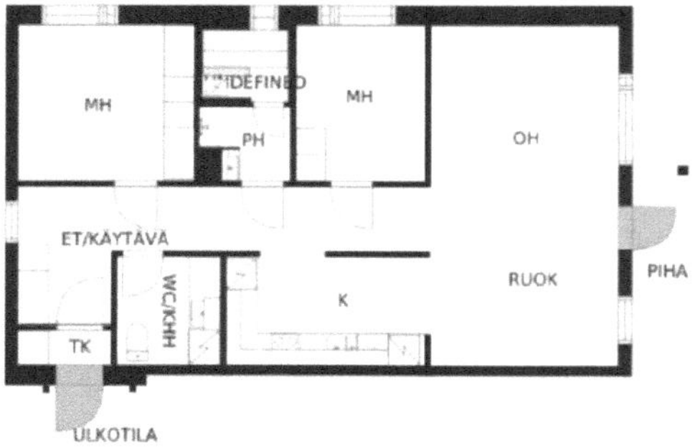

(b) Output Floor Plan with Highlighted Entrance Doors

Figure 12 : Indexed Masks of GMFS (GPT-integrated Multi-object Few-shot SAM)

Figure 13 : Semantic Classification of GMFS (GPT-integrated Multi-object Few-shot SAM

meanings with as few as five reference samples. The basic structure of GMFS includes five stages.

During the first stage, it recognizes similar regions in space while also learning the scale of different masks. The SAM encoder receives the test floor plan image and the five reference images, extracting and comparing visual characteristics from each image and generating the similarity map with higher values indicating more similarity. Since SAM generates object masks at varying scales it also must learn a scale parameter that helps inform how to combine the masks accordingly.

The second stage involves automatically extracting visual prompts instructing SAM on where to segment. From the similarity map, we filter out any areas with low similarity, group the remaining areas based on proximity, find the centroid of each group and use that as a point prompt, which enables SAM to assume the objects are spatially distributed to process multiple objects at one time.

The third stage generates the segmentation masks from the visual prompts; the masks were then checked for accuracy and only the masks that generated high confidence scores were kept. The processes involved in extracting and filtering masks were repeated in iterations to ultimately condense the output.

The fourth stage, mask post-processing, cleans the filtered masks more. Overlapping masks and very small fragments are discarded, and any void space within each segmented region is automatically filled. This adds a layer of refinement to reliable, complete masks.

In the fifth stage, GPT-4 is used to interpret the segmented masks, analyzing the functional purpose of each space. GPT-4 evaluates each area's function using its attributes around location, shape, size, spatial relationships of the masks, and gives it a label. In simple terms, each refined mask is indexed, and GPT-4 interprets the relevant function or assignment of each space or object based on user prompts.

This process has clear benefits compared to existing models. Conventional floor plan segmentation models, DeepFloorplan or U-Net, require pixel-by-pixel, manual annotation of the images, instructing humans to look at the drawing closely and assign each pixel in the drawing a value to generate masks. These practices are inefficient, overly labour-intensive, and, while automation makes sense in practice, impossible at scale for manual annotation. This has too many effects that obstruct models to increase their accuracy and generalise due to insufficient training data.

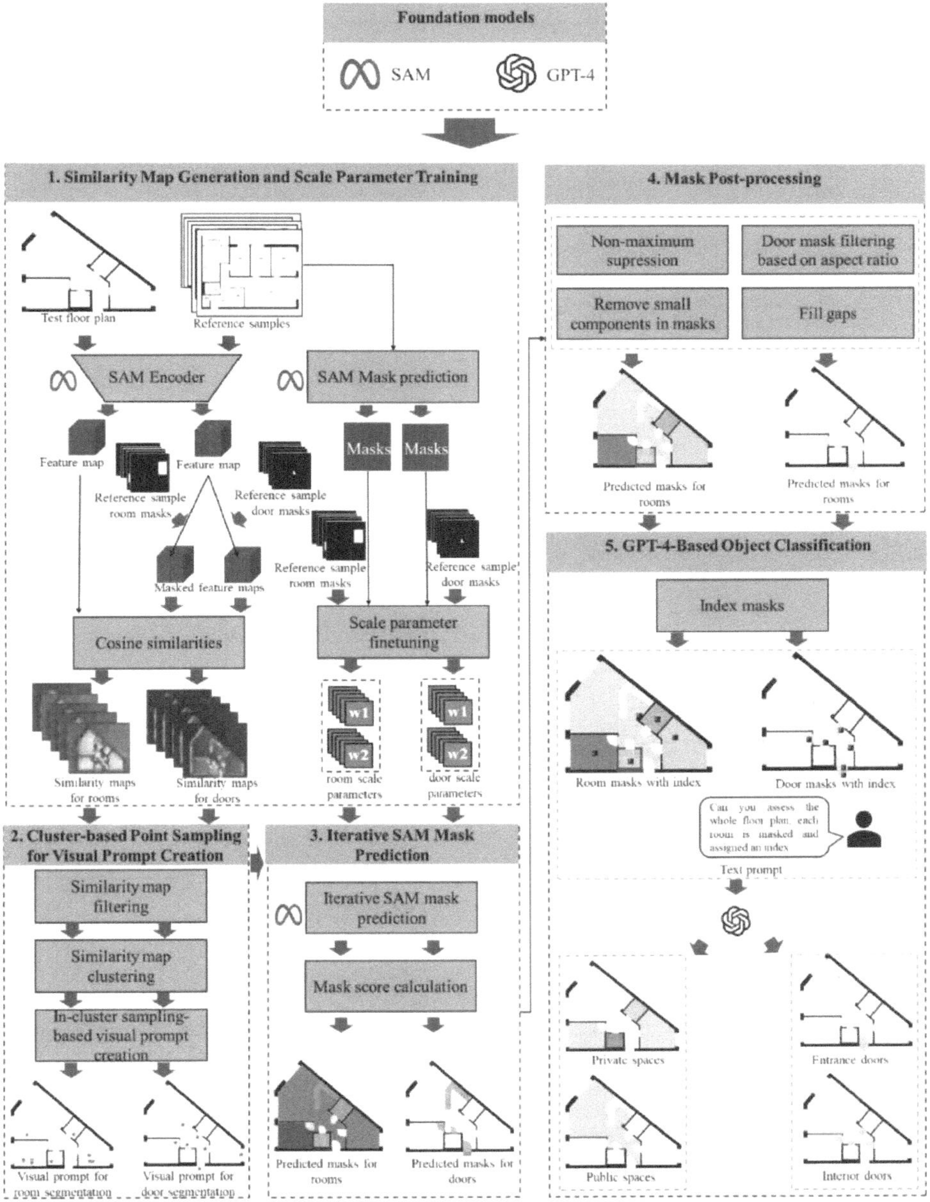

Figure 14 :
The overall architecture of GMFS (GPT-integrated Multi-object Few-shot SAM)

In contrast, a prompt-based interface allows users to simply click, draw bounding boxes, or use text instructions and the model can automatically mask and detect objects around those inputs. This will result in the fast and precise segmentation of hundreds of objects in less than a couple of minutes. More importantly, since it operates off a pre-trained foundation model, GMFS could easily be implemented to classify virtually any number of objects across diverse domains without the need for manual annotation thus allowing for large-scale dataset building to be feasible and efficient. In closing, our hybrid strategy-based SAM and ChatGPT arguably represent a primary solution for the limited training datasets issue as well as integrating semantics with visual analysis capabilities. This system can not only identify objects but understands their contextual framework structure while incorporating user-initiated instruction as a component of its operation.

Conclusion

For future work, there are multiple pathways to explore to enable further refinement within the hybrid system proposed throughout this study.
First, there is the initial task of creating a large-scale, high-quality dataset related specifically to architectural drawings. It is imperative that we get quite detailed and highly specialized annotations of many sizes and shapes of objects, particularly the irregularities of certain symbols and representations related to space, which we need to effectively train the models.

In addition to augmenting multimodal integration learning that would bring SAM's visual segmentation capabilities closely connected with ChatGPT's language reasoning capabilities. SAM takes images while ChatGPT takes text while each exists separately. Instead, future models could integrate information across the modalities allowing for not only object identification, but more valid higher-level analyses often used to analyze drawings or diagrams, like spatial logic, architectural codes and legalities, design intentions, etc. improvements that would allow language models, like GPT-4 to more fully engage with visual information present in drawings and diagrams. In this research, GPT-4 performed very well by understanding text but is not able to fully embrace visual spatial information or appropriate spatial information when interpreting and critiquing architectural drawings as drawn as included in architectural design. The importance of the model's spatial awareness and responding based on the inter-spatial relationships of architectural documents and plans will clearly be an important feature to include in the model in the future.
In addition, another challenge will be introducing the proposed platform into the architectural design world. The proposed platform will need to be implemented as an adjunct method to real-world design tools (e.g., BIM and CAD system) to facilitate daily architectural practices, including drawing analyses, code compliance, spatial simulation, with real-time interactivity, making the true potential as an architectural assistant.

Assuming sufficient future research is done to build system capabilities, interactivity will be another strong benefit. For example, being able to rapidly iterate and evaluate distinct designs in the early stages of an architectural project, the user would only need to input methods here - think of the time and human resources acquired.

In conclusion, a foundation-model based floor plan object recognition system is capable of benchmarking drawing assessments and has the potential for use as a planning system. With advanced research and technology trajectory, this approach may fundamentally change the game.

Figure

Reference

Among Zhiliang Zeng Xianzhi Li Ying Kin Yu Chi-Wing Fu (2020) '*Deep Floor Plan Recognition Using a Multi-Task Network with Room-Boundary-Guided Attention*',
The Chinese University of Hong Kong,27 February 2020, pp. 1-5.
Available at:
https://ieeexplore.ieee.org/document/9009528

Olaf Ronneberger, Philipp Fischer, and Thomas Brox (2015) '*U-Net: Convolutional Networks for Biomedical Image Segmentation*'
Computer Science Department and BIOSS Centre for Biological Signalling Studies, University of Freiburg, Germany, May 2015, pp. 1-2
Available at:
https://arxiv.org/abs/1505.04597

Nahian Siddique,Paheding Sidike, Colin Elkin and Vijay Devabhaktuni (2021), *U-Net and its variants for medical image segmentation: theory and applications*'
Department of Electrical and Computer Engineering, Purdue University Northwest, Department of Applied Computing, Michigan Technological University, June,2021, pp.1-12
Available at:
https://ieeexplore.ieee.org/document/9446143

Haolan Zhang, Ruichuan Zhang (2024), '*Few-shot learning with large foundation models for automated segmentation and accessibility analysis in architectural floor plans*'
Myers-Lawson School of Construction, Virginia Polytechnic Institute and State Univ., Blacksburg, VA, USA, December 2024, pp.1-16
Available at:
https://www.sciencedirect.com/science/article/pii/S2772991524000562

OpenAI (2024), *GPT-4 Technical Report*'
Mar.2024, pp.1-14
Available at:
https://cdn.openai.com/papers/gpt-4.pdf

Sebastien Bubeck, Varun Chandrasekaran, Ronen Eldan, Johannes Gehrke, Eric Horvitz, Ece Kamar, Peter Lee, Yin Tat Lee, Yuanzhi Li ,Scott Lundberg ,Harsha Nori, Hamid Palangi ,Marco Tulio Ribeiro, Yi Zhang (2023), '*Sparks of Arti cial General Intelligence:Early experiments with GPT-4* '
Microsoft Research, Apr.2023, pp.1-155
Available at:
https://www.researchgate.net/publication/369449949_Sparks_of_Artificial_General_Intelligence_Early_experiments_with_GPT-4

Alexander Kirillov, Eric Mintun2 Nikhila Ravi, Hanzi Mao2 Chloe Rolland, Laura Gustafson,

Tete Xiao, Spencer Whitehead Alexander C. Berg Wan-Yen Lo Piotr Doll´ar, Ross Girshick (2023) *Meta AI Research, FAIR, Apr.2024, pp.1-12*
Available at: https://arxiv.org/abs/2304.02643

Algorithmic Design in Healthcare System

A template and how-to guide for authors

Lucija Zupa[1]
[1] *Vienna University of Technology*
[1]*e12402346@student.tuwien.ac.at*

Nowadays, the efficiency of space usage is more important than ever. From an environmental point of view, but also from the point of different users where some of them spend around 30% of their adult life there. Healthcare is one of the biggest employers in the world but also gathers the most sensible kind of people and ones who need the most help. To help with easier wayfinding, financial cost and space organization to successfully separate patients from workers, algorithmic design became an indispensable tool. The possibility of arranging different departments and enabling simplicity of movement and going from point A to point B in the shortest and quickest possible route is the reason why these systems are becoming more and more needed. In this paper, we'll discuss different layout strategies and their main differences as well as pros and cons.

Keywords: *healthcare system, layout optimization, QAP, QSCP, MIP, ML, GCN*

INTRODUCTION

Algorithmic design is an architectural approach that involves using sequence of well-defined instructions used for solving problems and detecting which designing principle to implement. This approach has gained popularity since it enhances efficiency in planning.

When it comes to healthcare architecture, problems with finding solutions begin with endless potential ways to deal with it. Since one of the main foundations behind every good design is the amount of going through as many options as possible, here the idea is to develop a system which will do it for us. The idea of finding the optimal layout based on different algorithms provides the possibility to improve space based on their specific needs.

In this paper, the focus is on different systems that are already developed and some in use. We'll dive more into the details about their differences and the impact they have when used.

BACKROUND: FACILITY LAYOUT PROBLEMS IN HEALTHCARE

The architectural spacing design process shapes buildings, how they are used, and exploited. This process balances different fronts including but not limited to: aesthetics, efficiency, solar exposure, lightning, sustainability, etc.

Facility layout problem is part of architectural design and it involves careful consideration of possible space layouts to optimize its usage and efficiency. It is not only a consideration architects make, it also interests management science, engineering, and social science. While efficient layouts boost usage efficiency, they also increase users' satisfaction.

Primary objective for layout optimization is minimization of Material Handling Cost (MHC). In most general way, this includes not only cost of transportation of goods, but also of persons and other cost-affecting transports.

Facility layout optimization is unique for hospitals given their usage, need for inter-transportational efficiency, operational cost considerations, and patient-focused operation.

Hospitals are complex facilities and their spatial configuration affects patient experience, staff strain, and operational cost.

Traditional layouts can inadvertently lead to inefficiencies such as prolonged patient stay, elevated cost, or bottlenecks in critical pathways. For example, costs related to transportation within the medium-sized hospital account for almost $6 million as per [4].

Hospital Layout Problem (HLP) encompasses spatial allocation problem of hospital departments, rooms, and auxiliary areas. HLP shares domain with general Facility Layout Problems (FLPs) such as flow of occupants, emergency access, but also has unique considerations such as infection control, equipment placement, or regulatory constraints. Furthermore, [1] states that setup of a hospital directly affects the quality of services provided. Area accessibility and human interaction is also controlled by hospital layout. [2]

Transport optimization has proven to be the most challenging variable to optimize by human designers. Given many solutions and different factors that affect transportation cost, in addition to the other considerations that need to be taken. In large hospitals, more than 1,000 beds can be distributed over different departments. (Helber et al) These departments also offer training, and sometimes teaching to an associated university, as well as contain numerous administrative and social facilities. Within or between departments, significant flow of people and material occurs. In nature, transport is heterogeneous as some patients require different level of attention and help during transport, which in turn require different level of staff and space. Often, patient transport shares spaces with staff flow and material flow.

Beyond the transportation optimization, hospitals offer a challenge in optimization of other variables such as staff stand-by costs, energy efficiency, lightning, emergency-response time, etc.

It is obvious that a human designer is very unlikely to find the optimal point of such N-dimensional field. Such need contributed to rise in usage of algorithmic optimization methods as they are positioned to be a good tool for optimizations of such problems.

WHAT IS AN ALGORITMIC DESIGN

In general, algorithmic design can be defined as computer-aided design where a set of mathematic rules are used to deliver a result under user-defined conditions, boundaries and KPIs (Key Performance Indicators).

Algorithmic design was enabled by introduction of Computational Design in 1960s with milestones like first Sketchpad. [4] More and more advanced solutions followed and have profoundly changed the way architectural design is accomplished, streamlining object creation, visualization, flexibility, and design speed. Computer-aided design is not considered algorithmic design as it offers passive tool to designers to express their ideas and further elaborate them. Although as such, Computer-aided design is a tool of virtually every designer, it does not offer a tool for optimization or any kind of judging of design. As mentioned earlier, complex systems such as hospitals could be a good candidates if one is able to enhance Computer-aided Design tools with algorithmic design such that it can modify the design to control the results.

Beginnings of this idea were laid by Levin in [5] where one of first efforts to automate the design process were made. These efforts became more sophisticated over time where, for example, Wienzapfel [6] utilized geometry to define constraints and relationships for 3-D design. Medjdoub and Yannou [7] were first ones to use methods to generate dynamic spatial organization which followed by several researches and culminating with Wang and Zang [8] where they laid groundwork for facilitation of customization and automatic design of floor plans. Similarly, Arvin and House [9] created interactive and responsive design process. Finally, Nagi et al [10] introduced objective generic algorithm for offices, optimizing layouts while taking into consideration various benchmarks.

Algorithmic design for layout optimization problem can be solved using different methods such as Quadratic Assignment Problem (QAP), Mixed Integer Programming (MIP), graph theory, heuristic and metaheuristic approaches, and simulation-based models.

QAP was first introduces by Koopmans and Beckmann in 1957 [11] and it represents a difficult mathematical problem to solve algebraically. In addition, such problems are further complicated with real-world constraints. As such it is a good candidate for the algorithmic approach but it also has some downsides. Things such as equal-sized departments do not make it applicable to a hospital problem in its original form. Modified QAP, where departments are meshed into smaller grids does solve this limitation but makes the optimization exponentially more complex. The Quadratic Set Covering Problem (QSCP) offers an alternative for accommodating different department shapes [12]

MIP overcomes QAP limitations, and it uses several optimization models. For example, heuristic models provide quick, near perfect solutions but do not guarantee the most optimal solutions. Blend between these models and use of others to enhance such as AI is called metaheuristic approach.

Most recent approach to layout optimization utilizes Machine Learning (ML). Although AI was first time formulated in mid-20 century, only much later was it used for a first time in architectural domain on projects such as URBAN II or URBAN V. Even so, only recently has it evolved to reach a level of applicability in layout design. Breakthrough in architectural utilization was initiated when Generative Adversarial Networks (GANs) were introduced by Goodfellow [13] and followed by advancements of it such as CGAN, iGAN, GVM, and Pix2Pix. Researches by As et al [14], Peters [15], Newton [16], Wu et al [17], Chailou [18], Li et al [19], and Wang et al [20] showcase the versatility of depth of GAN applications in architecture.

Luo and Huang [21] introduced FloorplanGAN combining vector-based generation with raster-based discrimination for flat layout creation. Parallelly, the application of GNNs in architecture, Hu et al. [22] introduced Graph2Plan, a framework that uses Graph Neural Networks (GNNs) to automate architectural layout generation, improving design efficiency and flexibility by incorporating user-defined constraints. Liu et al. [23] addressed limitations in geometric detail capture and user customization by introducing a method that combines a Graph Convolutional Network (GCN) with an auto-regressive transformer to directly produce vector-based space layouts, marking a significant improvement in architectural design generation.

IMPLEMENTATION

Hospitals are important part of a society. Designing them in the most efficient way will significantly improve the quality of operation, patient satisfaction and costs. Early design stages of Hospital Layout Planning (HLP) are crucial as this is a point where layout and department locations are defined. Complexity and multifunctionality of hospitals demand multidisciplinary approach that integrates several stakeholders including architects, healthcare workers, and engineers.

Jia et al [xx] conducted a systematic review using spatial network analysis and simulation modeling and concluded the lack of clear methods of aligning design challenges with layout assessments. Jamali et al [xx] offered an architectural framework for ethically assessing layout issues. These contributions reveal the value algorithmic methods to improve layout efficiency.

Although facility layout planning (FLP) has its roots in manufacturing, its application within the healthcare sector is growing. [4] Still, research on utilization of such methods to hospitals is rare and usually comes form other branches rather than architecture. Techniques for implementing HLPs consist of three methods: procedural, algorithmic, and simulations. The lack of holistic approach means there is no comprehensive solution, incorporating all aspects of healthcare environments. Moreso, there is weight behind modeling optimization, mathematical programming, simulations, and lean manufacturing tools.

Regular way of hospital design usually uses Muther's Systematic Layout Planning (SLP) that is particularly because of its complete planning framework. In particular, "the relationship chart" is widely used for its effectiveness in optimizing spatial arrangements.

Algorithmic approaches use computer models to design or suggest layouts using user input such as objectives and constraints. These methods then rely on quantitative data for evaluation. [4] Although algorithmic models often need adjustments around detailed design aspects, they still take into consideration general aspects such as optimization objectives, legal constraints, and physical real-world. Cubukcuoglu [24] identifies the Quadratic Assignment Problem (QAP) and Mixed Integer Programming (MIP) as the leading algorithmic optimization strategies in hospital layout design. [24] Those methods are mainly used with a focus on minimizing patient travel a department intercommunication cost.

Simulation methods depict dynamic activities, patient movements, resource usage, and interactions. Jia et al. [97] emphasize that space syntax analysis (SSA) assesses layout designs in terms of visibility and access, while simulation modelling provides quantitative data on hospital operations. In addition to Simulation methods, there are also Spatial network analysis, Discrete event simulation, Agent-based modelling, and Machine Learning.

Machine learning usage is more extensive to residential buildings than to hospitals, nevertheless Zhao et al. [25] developed an AI method for emergency department (ED) layout design by first gathering a dataset of ED designs and then selecting over 100 high-quality layouts as training data for various Generative Adversarial Network. [4] Beyond that, Zhao et al. [26], and Zhou et al. [27] also used Machine Learning methods to optimize hospital layout but with limited success. While artificial intelligence does have a capability to optimize layouts in general, hospitals and their unique challenges, lack of data and constraints prevent an elegant solution applicable to those problems.

CASE STUDIES AND ANALYSIS

Further analysis focuses on specific case studies for Quadratic Assignment Problem (QAP) and Mixed Integer Programming (MIP).

Case study 1: Quadratic Assignment Problem (QAP)

Case study of Helber et al [27] uses QAP method on a real-world hospital as discussed above. Since authors use existing building to optimize material transportation costs, the building itself is used as a constraint.

The hospital under consideration here is Hannover hospital in Hannover, Germany. Besides being a hospital, it is also a leading medical school in Germany. The transportation data used in this study is adopted from [29]

Approach authors take is a two-stage approach. First stage is where Organizational Units (OU) are given their respective locations taking into consideration factors such as space requirements. In Stage II, internal optimization within each OU is executed.

Despite the problem being broken down to two phases, first one is still numerically challenging. Authors thus present "fix-and-optimize" heuristic approach to solve Phase I.

Authors state several factors that must be taken into consideration such as 1) space requirement per OU, 2) some OU need specific locations such as Emergency Room access to ambulance vehicle, and 3) some OU are prohibited to be next to each other while other are encouraged to be next to each other.

Furthermore, authors weight differently vertical and horizontal transport. Different types of transport are also introduced such as "emergency transport" that has to have maximum transport time.

At Stage I, authors apply same method as explained under "Implementation" of this article where each OU is broken down to several smaller peaces of it so QAP can optimize it to the constraints given by the building size and shape.

Stage II, in contrast to Stage I can be solved by standard programming software, although authors note that a solution may need to be subject to fine manual tuning by planners to consider additional aspects of the design.

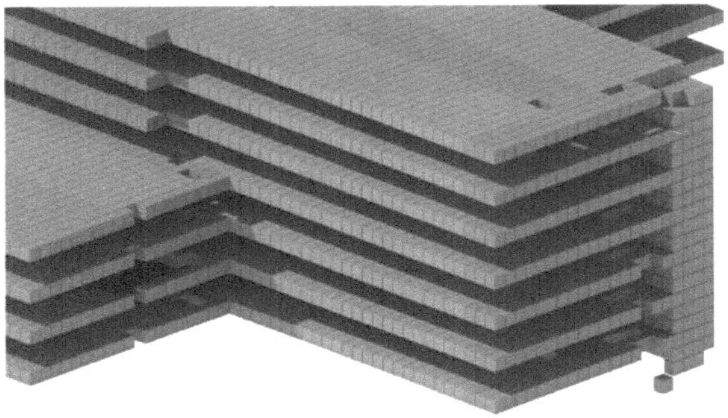

Figure 1
The walkable space as a mesh

Further, authors present "fix-and-optimize heuristic" model for Stage I problem. They eliminate the quadratic component of the optimization space by simply locating high-traffic OU closer to spatial center and other on a periphery. This reduces a problem to linear and is easily optimizable. Different approaches taken to eliminate quadratic components were to assign same weight to all KPIs, but this led to worse results than initial approach.

Researchers went further to run the same optimization algorithms for a virtual building for the new Hannover hospital and their latter approach outperformed approach former in two hours of computational time.

Case study 2: Study of optimization of Chinese hospital

Authors Xingren et al used Electronic Medical Records (EMRs) with sample size of 64,315 to produce systematic layout planning to understand patient flow.

Results met optimization objectives by reducing travel distance of general patients, high-volume department patients, and limited mobility patients by 5.84%, 4.58%, and 1.9% respectively.

Building of this study was 359 beds in Beijing, China from 1993.

Study has five steps; 1) Process mining, 2) Systemic layout planning, 3) Indoor optimal path planning, 4) Use of Generic Algorithm for optimization, and 5) Perform manual adjustments.

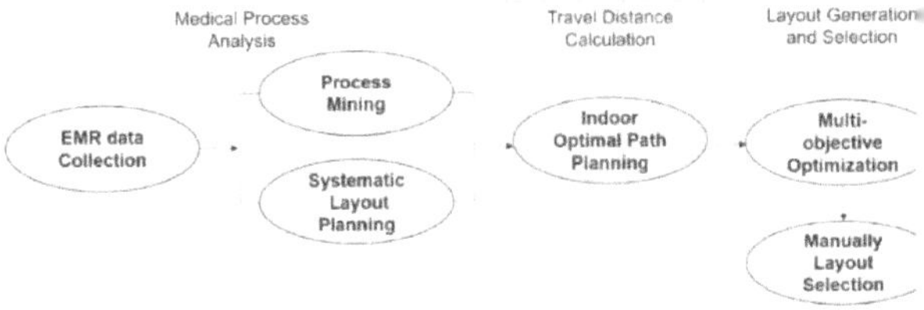

Figure 2
Data-driven framework

Data collection and Data analysis performed in this study include extraction of medical data from the database, taking into consideration any possible outliers such as New Year, time around lunch, student season, weekends, etc. Authors then processed this data to be useful for optimization such as extracting "To" and "From" travel data, grouping, and assigning difficulty.

Researchers assigned nodes to establish links within each floor and between the floors so that it presents all possible travel paths.

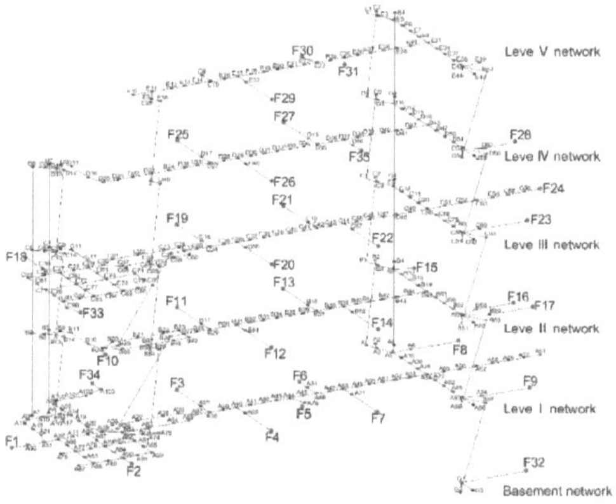

Figure 3
3D node network model

This network of paths, together with difficulty factor feeds to Generic Algorithm (GA). Authors go on to set specific weights to different type of travel (elevator v. stairs, etc.).

Constraints in this study were motivated by minimization of relocation expenses. Therefore, 1) only outpatient building was optimized, 2) non-patient areas were excluded, 3) high-relocation areas were excluded such as labs, diagnostic rooms, etc.

GA model used was Non-dominated Sorting Generic Algorithms II (NSGA-II) which is commonly used to ordering problems. Optimization was run in Python environment with both tournament selection and elitist selection. Before manually screening phase as a last step, the GA was run 18 times to provide many options for manual screening.

As with other optimization algorithms, manual selection needs to be performed as those algorithms do not take into consideration subtle design features and requirements. Final screening must comply with the following rules: 1) the selected layout should optimize the objectives of GA, 2) it should comply with building code, 3) it should ensure ideal functional continuity including patients' wayfinding, and 4) layouts with minor adjustments are prioritized. Final selection was taken upon 215 pareto solutions from GA model.

CONCLUSION

The implementation of hospital layout optimization helps significantly with a more data-driven and systematic approach in solving more complex problems. To summarize, the comparison of QAP and MIP provides us with an overview of two used algorithmic programs. When it comes to QAP which is known as a foundational method for special organization, it seems it is not an appropriate tool when it comes to hospitals. Although MIP has overcome the QAP limitations, has more flexibility, it doesn't guarantee the most optimal solution. However, new algorithms such as GANs and GCNs are showing signs in improving those limitations as well. Overall, these tools are becoming more and more integrated in the designing process to help achieve patient-friendly environment and adapt to future needs.

REFERENCES

[1] R.S. Ulrich, C. Zimring, X. Zhu, J. DuBose, H.-B. Seo, Y.-S. Choi, X. Quan, A. Joseph, A review of the research literature on evidence-based healthcare design, HERD 1 (3) (2008) 61, https://doi.org/10.1177/193758670800100306

[2] C. Hicks, T. McGovern, G. Prior, I. Smith, Applying lean principles to the design of healthcare facilities, Int. J. Prod. Econ. 170 (2015) 677–686, https://doi.org/10.1016/j.ijpe.2015.05.029

[3] Aysegul Ozlem Bayraktar Sari, Wassim Jabi, Architectural spatial layout design for hospitals: A review, Journal of Building Engineering, Volume 97, 2024, 110835, ISSN 2352-7102, https://doi.org/10.1016/j.jobe.2024.110835

[4] MODELING AND ANALYSIS OF THE HOSPITAL FACILITY LAYOUT PROBLEM by Amol Shrikrishna Padgaonkar

[5] P.H. Levin, Use of Graphs to Decide the Optimum Layout of Buildings, Building Research Station, 1964.

[6] G. Weinzapfel, T.E. Johnson, J. Perkins, IMAGE: an interactive computer system for multi-constrained spatial synthesis. Proceedings of the 8th Design Automation Workshop, 1971, pp. 101–108, https://doi.org/10.1145/800158.805064

[7] B. Medjdoub, B. Yannou, Dynamic space ordering at a topological level in space planning, Artif. Intell. Eng. 15 (1) (2001) 47–60, https://doi.org/10.1016/S0954-1810(00)00027-3

[8] X.-Y. Wang, K. Zhang, Generating layout designs from high-level specifications, Autom. ConStruct. 119 (2020) 103288, https://doi.org/10.1016/j.autcon.2020.103288

[9] S.A. Arvin, D.H. House, Making designs come alive: using physically based modeling techniques in space layout planning, in: G. Augenbroe, C. Eastman (Eds.), Computers in Building: Proceedings of the CAADfutures'99 Conference. Proceedings of the Eighth International Conference on Computer Aided Architectural Design Futures Held at Georgia Institute of Technology, Atlanta, Georgia, USA on June 7–8, 1999, Springer US, 1999, pp. 245–262, https://doi.org/10.1007/978-1-4615-5047-1_16

[10] D. Nagy, D. Lau, J. Locke, J. Stoddart, L. Villaggi, R. Wang, D. Zhao, D. Benjamin, Project discover: an application of generative design for architectural space planning, https://doi.org/10.22360/simaud.2017.simaud.007, 2017

[11] T.C. Koopmans, M. Beckmann, Assignment problems and the location of economic activities, Econometrica 25 (1) (1957) 53–76, https://doi.org/10.2307/1907742

[12] M.S. Bazaraa, Computerized layout design: a branch and bound approach, A I I E Transactions 7 (4) (1975) 432–438, https://doi.org/10.1080/05695557508975028

[13] I.J. Goodfellow, J. Pouget-Abadie, M. Mirza, B. Xu, D. Warde-Farley, S. Ozair, A.C. Courville, Y. Bengio, Generative adversarial nets. Neural Information Processing Systems, 27, 2014, pp. 2672–2680 Red Hook, NY.

[14] I. As, S. Pal, P. Basu, Artificial intelligence in architecture: generating conceptual design via deep learning, Int. J. Architect. Comput. 16 (4) (2018) 306–327, https://doi.org/10.1177/1478077118800982.
[15] N. Peters, *Enabling Alternative Architectures : Collaborative Frameworks For Participatory Design* [Master's Thesis, Harvard University Graduate School of Design], 2018.
[16] D. Newton, Deep generative learning for the generation and analysis of architectural plans with small datasets, https://doi.org/10.5151/proceedingsecaadesigradi2019_135, 2019.
[17] W. Wu, X. Fu, R. Tang, Y. Wang, Y. Qi, L. Liu, Data-driven interior plan generation for residential buildings, ACM Trans. Graph. 38 (2019) 1–12.
[18] M. Li, A.G. Patil, K. Xu, S. Chaudhuri, O. Khan, A. Shamir, C. Tu, B. Chen, D. Cohen-Or, H. Zhang, GRAINS: generative recursive autoencoders for Indoor scenes, ACM Trans. Graph. 38 (2) (2019) https://doi.org/10.1145/3303766 , Article 12
[19] K. Wang, Y.-A. Lin, B. Weissmann, M. Savva, A.X. Chang, D. Ritchie, PlanIT: planning and instantiating indoor scenes with relation graph and spatial prior networks, ACM Trans. Graph. 38 (4) (2019) 1–15, https://doi.org/10.1145/3306346.3322941
[20] S. Kim, S. Park, H. Kim, K. Yu, Deep floor plan analysis for complicated drawings based on style transfer, J. Comput. Civ. Eng. 35 (2) (2021), https://doi.org/10.1061/(ASCE)CP.1943-5487.0000942
[21] Z. Luo, W. Huang, FloorplanGAN: vector residential floorplan adversarial generation, Autom. ConStruct. 142 (2022) 104470, https://doi.org/10.1016/j.autcon.2022.104470
[22] N. Nauata, K.-H. Chang, C.-Y. Cheng, G. Mori, Y. Furukawa, House-GAN: relational generative adversarial networks for graph-constrained house layout generation, European Conference on Computer Vision 162 (2020) 177, https://doi.org/10.1007/978-3-030-58452-8_10
[23] J. Liu, Y. Xue, J. Duarte, K. Shekhawat, Z. Zhou, X. Huang, End-to-end graph-constrained vectorized floorplan generation with panoptic refinement. European Conference on Computer Vision, 2022, pp. 547–562
[24] C. Cubukcuoglu, HOPCA: Hospital Layout Design Optimization Using Computational Architecture, Doctoral thesis, Delft University of Technology, 2023 https://doi.org/10.7480/abe.2023.03, Publication Number
[25] C. Zhao, J. Yang, J. Li, Generation of hospital emergency department layouts based on generative adversarial networks, J. Build. Eng. 43 (2021) 102539, https://doi.org/10.1016/j.jobe.2021.102539
[26] C. Zhao, J. Yang, W. Xiong, J. Li, Two generative design methods of hospital operating department layouts based on healthcare systematic layout planning and generative adversarial network, J. Shanghai Jiaot. Univ. 26 (1) (2021) 103–115, https://doi.org/10.1007/s12204-021-2265-9.
[27] Y. Zhou, Y. Wang, C. Li, L. Ding, C. Wang, Automatic generative design and optimization of hospital building layouts in consideration of public health emergency, Eng. Construct. Architect. Manag. 31 (4) (2022) 1391–1407, https://doi.org/10.1108/ECAM-08-2022-0757
[28] Stefan Helber, Daniel Bo¨hme, Farid Oucherif, Svenja Lagershausen, Steffen Kasper "A hierarchical facility layout planning approach
for large and complex hospitals" Published online: 25 February 2015 Springer Science+Business Media New York 2015
[29] Oucherif F (2012) Analyse und Optimierung der Layoutplanung von

Krankenhäusern aus Sicht der Transportlogistik durch Methoden des Operations Research. Gottfried Wilhelm Leibniz UniversitätHannover, Institut für Produktionswirtschaft, Masterarbeit

Machine Learning Algorithms Supporting Life Cycle Assessment: A Review

Kinga Średzińska
Technische Universität Wien
e12410042@student.tuwien.ac.at

Life Cycle Assessment (LCA) is a systematic methodology for evaluating the environmental impacts of products and processes. As LCA studies become more complex, challenges such as data gaps, modeling uncertainty, and computational demands have emerged. In recent years, machine learning (ML) algorithms have been increasingly adopted to address these challenges, offering improved data handling, prediction accuracy, and automation.

Keywords: *Life Cycle Assessment, Machine Learning, Environmental Impact, Prediction, Model selection.*

INTRODUCTION

Life Cycle Assessment (LCA) is a methodological framework for evaluating the environmental impacts associated with all stages of a product's life cycle. Traditional LCA practices often face limitations due to restricted data availability, reliance on static emission factors, and the time-consuming nature of manual modeling. Advancements in data science have led to increased interest in integrating machine learning (ML) techniques into LCA. While studies suggest that ML can enhance predictive accuracy and automate the assessment process, the integration of ML into LCA lacks standardized methodologies.

This paper reviews the recent literature and case studies focusing on key studies to highlight applications, benefits, and ongoing challenges of integration ML algorithms into LCA.

MACHINE LEARNING IN LIFE CYCLE ASSESSMENT

Machine learning (ML) is increasingly utilized in life cycle assessment (LCA) to enhance prediction accuracy, automate data processing, and manage complex datasets. (Romeiko et al., 2024; Ahmed et al., 2023) ML techniques allow for quicker environmental impact assessments across various sectors. (Romeiko et al., 2024; Cha & Park, 2025)

Motivation for ML-assisted LCA

Traditional LCA methods often rely on static emission factors and averages that fail to capture the variability in real-world systems, as noted by Cha & Park (2025). This limitation is further illustrated by Ahmed et al. (2023), who demonstrate that emissions from building demolition can vary substantially depending on specific process parameters, equipment used, and material selection. These challenges impact the accuracy, scalability, and responsiveness of conventional LCA approaches. However, integrating machine learning (ML) techniques can help address these methodological gaps and foster more robust, data-driven assessments. (Cha & Park, 2025; Romeiko et al., 2024)

Types of ML Algorithms Used in LCA

A range of machine learning algorithms have been adopted in LCA research to meet diverse analytical needs. Supervised learning methods, such as gradient boosting machines, decision trees, random forests, and neural networks, are widely used for predicting environmental impacts based on process parameters. Supervised learning method, including gradient boosting machines, decision trees, random forests, and neural networks, are commonly employed to predict environmental impacts based on process parameters (Cha & Park, 2025; Romeiko et al., 2024). Furthermore, Romeiko et al. (2024) emphasize that unsupervised learning techniques, such as clustering and dimensionality reduction, can help identify patterns and drivers within extensive and complex LCA datasets. Additionally, reinforcement learning has the potential to aid process optimization and dynamic scenario analysis by simulating adaptive decision-making in response to environmental feedback (Romeiko et al., 2024). For instance, in assessing building demolition, Cha & Park (2025) compared various supervised ML algorithms, including artificial neural networks, decision trees, random forests, and gradient boosting machines. They found that the gradient boosting machine (GBM) model exhibited the highest predictive performance, showcasing the model's strong generalization capability and accuracy in estimating CO_2 emissions based on specific features of demolition projects.

KEY APPLICATIONS OF MACHINE LEARNING IN LCA

Key applications of machine learning in LCA, concluded by Romeiko et al. (2004), include predicting life cycle inventory and impact assessment results, filling data gaps, and automating the analysis of complex systems.

Data Collection and Preprocessing

Successful application of ML in LCA depends on the availability of high-quality data and thorough preprocessing. (Ahmed et al., 2023; Cha & Park, 2025) Ahmed et al. (2023) developed a comprehensive dataset focused on additive manufacturing processes, while Cha and Park (2025) assembled detailed data covering 186 building demolition projects, including variables such as structure type, floor area, equipment used, and energy consumption. To ensure data reliability, common preprocessing steps were employed across studies, including handling missing values, removing outliers, and standardizing inputs. These procedures have become essential best practices in ML-enhanced LCA research, supporting more accurate and robust model training and evaluation. (Romeiko et al., 2024)

Feature Engineering and Model Training

To optimize model accuracy Cha & Park (2025) emphasized role of feature engineering (e.g. selecting critical variables) and hyperparameter tuning, which is a methodological practice now widely adopted in ML-assisted LCA to refine predictive performance. This approach underscores the trend in the field of prioritizing variable significance and algorithmic optimization to enhance reliability, as demonstrated in additive manufacturing and literature reviews. (Ahmed et al., 2023; Romeiko et al., 2024)

Model Evaluation and Interpretation

Model evaluation in ML-based LCA commonly uses R^2, which shows how well predictions match actual data, and mean absolute error (MAE), which measures average prediction error. High R^2 and low MAE indicate strong model performance. For instance, Cha & Park (2025) reported R^2 values of 0.997 for training, 0.983 for testing, and 0.984 for validation, demonstrating that their machine model provided highly accurate and generalizable predictions. Similarly, Ahmed et al. (2023) found that integrating ML into additive manufacturing LCA significantly improved both prediction accuracy and computational efficiency compared to manual methods.

BENEFITS OF ML-ASSISTED LCA

The integration of machine learning into LCA enhances the accuracy of environmental impact predictions, facilitating informed decision-making during the design and planning phases. (Romeiko et al., 2024; Cha & Park, 2025) As noted by Cha and Park (2025), the capability to forecast demolition-related CO_2 emissions aids in optimizing building designs and implementing effective carbon reduction strategies. Additionally, in additive manufacturing, ML models contribute to real-time decision-making and sustainability optimization (Ahmed et al., 2023).

Romeiko et al. (2024) and Cha & Park (2025) emphasized that ML also improves the life cycle inventory and impact assessment by estimating missing or uncertain data, which enhances the reliability of inventories, particularly in data-scarce stages such as demolition and manufacturing.

Furthermore, the emergence of artificial intelligence (AI) and large language models (LLMs) raises concerns regarding their environmental impact. To address this, Fernandez et al. (2024) proposed a holistic LCA framework that employs ML methods to analyze power use and emissions throughout the development and deployment of LLMs, highlighting how ML can promote sustainable design and operation of digital technologies.

Compared to traditional LCA methods, ML approaches provide notable advantages in scalability and efficiency. According to Cha & Park (2025) automated data processing and model training reduces manual labor involved in data collection and analysis. That allows for the rapid assessment of multiple scenarios or products. (Cha & Park, 2025)

CHALLENGES FOR ML-ASSISTED LCA

According to Romeiko et al. (2024) the integration of machine learning into LCA presents significant challenges, particularly in data quality and availability. A key issue is the scarcity of large, high-quality datasets necessary for effective analysis. Yet, Cha & Park (2025) addressed this limitation in the demolition sector by developing a comprehensive dataset. That emphasizes the need and possibility for similar solutions across different sectors and life cycle stages.

Another critical challenge lies in model transparency and interpretability. Many ML algorithms, particularly deep learning models, often are complicating the interpretation and validation of their predictions. (Romeiko et al., 2024) However, techniques such as feature importance analysis, as demonstrated by Cha & Park (2025) can provide insights into the variables that influence model outcomes.

TRENDS AND FUTURE RESEARCH

Future research in machine learning-assisted life cycle assessment is moving toward several directions.

Ahmed et al. (2023) and Romeiko et al. (2024) observe that hybrid approaches that combine machine learning with traditional LCA methodologies and domain expertise show potential to enhance both the accuracy and interpretability of results. This integration ensures that nuanced domain knowledge reduces the risk of misrepresentation. (Ahmed et al., 2023)

Dynamic and real-time LCA is another emerging frontier enabled by machine learning as highlight Fernandez et al. (2024). ML algorithms can continuously process and integrate new data streams, allowing LCA results to be updated in real time as product designs, supply chains, or operational conditions change. Real-time LCA is also reported to enhance its responsiveness. That as a result enables organizations to quickly identify inefficiencies, optimize environmental performance, and adapt to new sustainability challenges. (Fernandez et al. 2024)

CONCLUSIONS

Machine learning is improving data processing, prediction accuracy, and model automation of traditional life cycle assessment processes. Thus, the integration with ML helps address key challenges in LCA, such as data gaps, limited scalability, and time-intensive analysis.

ML offers strong potential to enhance LCA's effectiveness and expand its role in sustainability analysis across diverse industries. In sectors such as construction, manufacturing, and digital technologies, ML enhances the estimation of CO_2 emissions, supports the completion of life cycle inventories, and enables more precise interpretation of environmental impacts. Its ability to quickly fill data gaps and extract insights from complex datasets makes ML a valuable tool for informed sustainability planning and decision-making.

However, challenges around data quality, model transparency, and the need for standardized methods remain. The development of standardized protocols for model integration and reporting is increasingly recognized as essential for further research.

REFERENCES

Ahmed, N.Z., Defersha, F., Xu, X. and Yang, S. (2023) 'Automating life cycle assessment for additive manufacturing with machine learning: Framework design, dataset buildup, and a case study', *Journal of Cleaner Production*, 412.

Cha, G.-W. and Park, C.-W. (2025) 'Development of an optimal machine learning model to predict CO_2 emissions at the building demolition stage', *Buildings*, 15, 526. https://doi.org/10.3390/buildings15040526

Fernandez, J., Na, C., Bisk, Y. and Strubell, E. (2024) 'Evaluating the Environmental Impact of Language Models with Life Cycle Assessment', *Proceedings of the 2024 AAAI Conference on Artificial Intelligence*.

Romeiko, X.X., Zhang, X., Pang, Y., Gao, F., Xu, M. and Lin, S. (2024) 'A review of machine learning applications in life cycle assessment studies', *Science of the Total Environment*, 912.

A review of Machine Learning-Based Automated Fault Detection and Diagnostics Methods in HVAC Systems

Ellen Destoop
Technical University of Vienna
e12420000@student.tuwien.ac.at

Automated Fault Detection and Diagnostics (AFDD) in HVAC building systems has become more data-driven through Machine Learning (ML). This paper presents a review of different recent contributions to ML-based AFDD, each addressing different approaches and challenges, such as false alarm reduction, limited labelled data and system and building generalizability. This review aims to reflect the recent evolution of ML-based AFDD and summarizes the approaches across percentile-based residual scoring, anomaly detection and unsupervised- and hybrid models.

Keywords: *Machine Learning, Automated Fault Detection, Building Systems, HVAC.*

INTRODUCTION

Heating, Ventilation, and Air Conditioning (HVAC) systems are playing an important role in ensuring indoor environmental quality and energy efficiency in buildings (Shi and O'Brien, 2019). These systems are becoming more complex and are therefore prone to a wide range of faults. These can significantly increase energy consumption and have a big impact on occupant comfort (Babadi Soultanzadeh *et al.*, 2025). As a result, Automated Fault Detection and Diagnostics (AFDD) has become important in the research of modern building management systems.

Traditional AFDD methods, such as rule-based logic or first-principles modelling, are easy to implement but are often not easily scalable and rely on knowledge from experts. In contrast, machine learning offers data-driven adaptive solutions with the help of improved data and computing tools.

This paper reviews four recent contributions to the field of AFDD for building systems that present different solution strategies. First, the paper by Shi and O'Brien (2019) provides a structured framework for developing and implementing AFDD systems. It reviews and compares a diverse range of methods for future generation, fault detection and fault diagnostics. Secondly, the method by Dietz et al. (2024) proposes a percentile-based scoring system with a dynamic tolerance band for residual evaluation to improve interpretation and sensitivities of faults.

Babadi Soultanzadeh et al. (2025) introduces an unsupervised AFDD framework for light commercial buildings. Their approach is designed to work with unlabelled and incomplete data and is validated by data from two buildings with different data quality. Lastly, A real-world implementation is presented by Nelson and Dieckert (2024), who combines decision trees and anomaly detection algorithms in a large university building.

This review presents their contributions, compares their approaches, and identifies trends, challenges, and opportunities. The goal is to provide a clear understanding of how machine learning techniques are transforming automated fault detection in building HVAC systems.

FUNDAMENTALS OF AFDD

Automated Fault Detection and Diagnostics is a process that monitors and identifies abnormal behaviour in various building systems. These systems, including HVAC systems are becoming more complex, dynamic and need to operate under varying conditions. AFDD helps to ensures safety, efficiency, and reliability by detecting faults before they develop into failures. A fault refers to a deviation from a certain range of measured values, which can also be called a residual. This can occur from many various causes and since modern building systems are more and more interconnected, one fault in a component can affect others (Shi and O'Brien, 2019).

Detection of these faults is the process of identifying when something is not working as intended. It compares observed values collected by sensors to expected values and reports significant differences. This can be obtained by using static or dynamic thresholds, statistical techniques, or complex decision models. In a traditional system, fault detection is manually done by technicians, while AFDD automates this process.

When a fault is detected, it is further diagnosed to find out what the fault is and where it happened, which can be achieved with the help of analytical or rule-based models. These systems also have their limitations, such as high costs for set up and maintenance and it is harder to generalize across buildings since it is often tailored to a system. To address these challenges, research on machine learning approaches for AFDD picked up and became central in the development of modern AFDD solutions. Nowadays methods range from supervised classifiers, which are trained to recognize specific faults to unsupervised anomaly detectors that can identify a fault without labelled fault data (Shi and O'Brien, 2019).

APPROACHES OF ML-BASED AFDD

The paper by Dietz et al. (2024) proposes a new method that focuses on how to better interpret results or residuals of machine learning models that predict how HVAC systems should be operating. Traditional residual evaluation has different limitations such as frequent false alarms due to fixed thresholds not corresponding to behaviour changes of HVAC systems.

Dietz et al. developed a dynamic and more intelligent method that trains ML models to predict normal values for the system and compares them to actual values. They use percentiles to decide how big the difference between those two values can be before it is considered irregular. Each difference or residual is given a score between 0 (normal) and 1 (faulty). This scoring is further adjusted based on the operating of the system. If the sum of all scores of all sensors is too high for a certain time-step, it will be flagged as a fault. The researchers trained and tested the ML models with real data of nine air handling units in 3 different buildings. As a result, their dynamic method in comparison to a traditional static method had fewer false alarms. It performed well across different buildings, and it showed good flexibility (Dietz, Scholzen and Réhault, 2024).

The paper by Babadi Soultanzadeh et al. (2025) presents an unsupervised ML approach. Since labelled fault data is often limited, they developed a method that does not require labelling or expert knowledge. It analyses patterns directly from raw data and calculates reconstruction errors through a principal component analysis (PCA). This method can identify when the system doesn't behave normal. It also uses time-windowing to better understand how faults develop over time.

To improve accuracy, the authors applied data cleaning and subset selection. The method was tested on two light commercial buildings with different HVAC systems and showed a reliable detection of faults, even with incomplete, noisy, or unlabelled data. By using standard Python tools, this computational method is well suited for practical implementation (Babadi Soultanzadeh *et al.*, 2025).

Nelson and Dieckert (2024) designed and tested a AFDD approach for HVAC systems in a university building that combines machine learning and traditional rule-based methods. They used a decision tree-based classification algorithm, called LightGBM, to detect if it is a normal or faulty operation. Furthermore, they use principal component analysis to detect anomalies in the system data.

Their method successfully detected immediate faults and more gradual issues. Compared to standard rule-based checks, the ML models required less manual configuration and were more adaptable. The results are also easier to understand, due to the high interpretability of these decision trees. The anomaly detection methods were useful, but sometimes a manual follow-up was needed to confirm the nature of the fault (Nelson and Dieckert, 2024).

DISCUSSION

The methods presented in the selected papers differ in terms of methodological structure, data requirements, supervision, and level of interpretability. Table 1 summarizes these differences and similarities.

Table 1
Comparative analysis

Study	ML method	Model(s)	Supervision	Interpretability
Dietz et al. (2024)	Residual scoring	Percentile-based scoring	Model-based	Residual scoring
Babadi Soultanzadeh et al. (2025)	Anomaly detection	Principal component analysis	Unsupervised	Squared Prediction Error
Nelson and Dieckert (2024)	Classification+ anomaly detection	Principal component analysis+ LightGBM	Hybrid	Decision tree

One of the main challenges in applying machine learning to AFDD is the lack of labelled fault data. This makes unsupervised methods such as those by Dietz et al. (2024) and Babadi Soultanzadeh et al. (2025) relevant as they can detect anomalies from raw sensor data. This is practical in real-world settings where data can be incomplete or noisy. When labelled data is available, supervised models show a great precision. The hybrid approach, developed by Nelson and Dieckert (2024), combines LightGBM classification with PCA anomaly detection and performs better than simple rule-based systems.

Interpretability is also an important factor in AFDD system design. The presented methods offer varying degrees of interpretability:

- Percentile-based scoring translates residuals into clear scores based on statistical norms.
- Decision trees offer logic structures that are similar to traditional Boolean rules.
- PCA-based methods provide indicators like Squared Prediction Error, which can be interpreted by experts.

Across all studies, the ability to detect abrupt faults is proved. The detection of gradual faults happening over time is also addressed. Adaptive thresholds and baseline modelling, used by Dietz et al. (2024) and Babadi Soultanzadeh et al. (2025), help catch these early. Nelson and Dieckert's (2024) hybrid model also detects developing faults by spotting unusual trends and patterns in the data.

CONCLUSION

This review compared three recent ML-based approaches to automated fault detection and diagnostics (AFDD) in building HVAC systems. The selected methods reflect a shift from rule-based logic toward data-driven, are better scalable across different systems and buildings, and offer interpretable solutions. The studies present a range of methodologies, including dynamic residual scoring, unsupervised anomaly detection using principal component analysis, hybrid supervised-unsupervised models and conceptual frameworks for AFDD system design.

Strengths such as the ability to detect both abrupt and gradual faults, and better interpretability with residual scoring and decision tree logic were identified. Furthermore, unsupervised and hybrid approaches showed potential in real-world integration due to the possibility to use unlabelled data, which is a common limitation. There are still challenges, especially in automating diagnostics and standardizing evaluation datasets and thresholds. As building systems grow in complexity and the availability of data further increases, the role of machine learning in AFDD is expected to grow. Further work should focus on these elements, closing the gap between detection and diagnostics, and on the integration of AFDD systems into building management flows.

REFERENCES

Babadi Soultanzadeh, M. *et al.* (2025) 'Unsupervised automated fault detection and diagnosis for light commercial buildings' HVAC systems', *Building and Environment*, 267, p. 112312. Available at: https://doi.org/10.1016/j.buildenv.2024.112312. (Accessed 1 June 2025)

Dietz, S., Scholzen, F. and Réhault, N. (2024) 'Enhancing automated fault detection in building systems: A percentile-based scoring approach with dynamic tolerance range for residual evaluation', in. *BauSim Conference 2024.* Available at: https://doi.org/10.26868/29761662.2024.44. (Accessed 25 May 2025)

Nelson, W. and Dieckert, C. (2024) 'Machine Learning-Based Automated Fault Detection and Diagnostics in Building Systems', *Energies*, 17(2), p. 529. Available at: https://doi.org/10.3390/en17020529. (Accessed 3 June 2025)

Shi, Z. and O'Brien, W. (2019) 'Development and implementation of automated fault detection and diagnostics for building systems: A review', *Automation in Construction*, 104, pp. 215–229. Available at: https://doi.org/10.1016/j.autcon.2019.04.002. (Accessed 1 June 2025)

Evaluating Generic Algorithms for Early-Stage Building Design Optimization

Sinan İlker[1],
[1,] *Technische Universität Wien*
[1] *e12432842@student.tuwien.ac.at*

Genetic Algorithms (GAs) are widely recognized as effective search in early-phase architectural design, where the complexity and openness of the design space challenge traditional methods rather than complex. This paper investigates the application of GAs for performance-driven optimization in the conceptual design stage, drawing from recent advancements in the field.

Keywords: *Genetic Algortihms, Design Optimization, Early-Stage Architecture, Evolutionary Computation, Parametric Design, LLM's*

INTRODUCTION

In architectural design, the early planning phase plays a decisive role in shaping both spatial quality and performance outcomes. At this stage, decisions are often made based on intuition or simplified heuristics, despite the significant long-term impact they have on energy use, daylight access, material efficiency, and user comfort (Nazari and Yan, 2025). As the complexity of design problems continues to grow, algorithmic approaches have been increasingly adopted to support informed and explorative decision-making. Among these, Genetic Algorithms (GAs) have become particularly valuable tools due to their capacity to search through large, multidimensional solution spaces without requiring derivatives or strict constraints (Deb, 2001). Inspired by principles of natural selection, GAs offer a flexible optimization method that evolves design alternatives through iterative cycles of selection, crossover, and mutation. This makes them highly suitable for problems in architecture where performance goals may be competing, loosely defined, or difficult to quantify early on. Among these, Genetic Algorithms (GAs) have become particularly valuable tools due to their capacity to search through large data, multidimensional solutions spaces without requiring alternatives or strict constraints (Deb, 2001). Getting inspired by principles of natural selection, GAs offer a flexible optimization method that evolves design alternatives through different cycles of selection, crossover, and mutations overall. This makes them highly suitable for problems in architecture where performance goals may be competing, loosely defined, or difficult to quantify early on for the optimization and the design progress. Recent developments in visual scripting environments and programs, particularly in Grasshopper and similar tools, have made the integration of evolutionary solvers more accessible to designers within a range of different approaches. Tools like Galapagos and Octopus allow users to define fitness functions and observe real-time optimization processes in a visual and intuitive way (Wilensky, 1999; Lee, 2010). These tools extend the designer's capacity to test multiple scenarios and encourage a more experimental and performance-oriented design culture on the overall Optimized Design approaches overall. Despite their strengths, GAs come with limitations. They may require a high number of generations to supply, which can be computationally demanding for the current approaches. Furthermore, the interpretability of the evolved solutions may not always align with architectural judgment or constructability of the designs. This paper investigates both the potential and challenges of applying Genetic Algorithms in early-phase design optimization and offers reflections based on literature and tool-based examples.

ALGORITHMIC APPROACH AND APPLICATION CONTEXT

Genetic Algorithms are problem-solving methods that work by mimicking natural evolution in the phase. They are useful in architectural design because they can search through many possible solutions, even when the design problem is complex or hard to define clearly. Very Unlikely traditional methods that need precise rules or smooth changes, Genetic Algorithms can handle rough, irregular design spaces, which are common in early stages of planning (Deb, 2001).

The basic idea of a Genetic Algorithm is simple: you start with a group of random solutions in the approach, and check how good each one is using a fitness score on the board, and then keep the better ones to create new solutions while still looking for the best optimized one under the different parameters. These new solutions are made by combining parts of the old ones and making small changes, similar to biological crossover and mutation in the nature. Over time, the solutions usually get better. This makes Genetic Algorithms a good way to explore different options without needing to know the best answer from the start (Bentley, 1999).

In architecture, these algorithms are often used inside tools like Grasshopper for Rhino or Dynamo for Revit. Galapagos is a common example for single goals, and Octopus is used when there are multiple goals. Designers can connect these tools to simulation software like Ladybug (with Honeybee) or EnergyPlus to check things like daylight or energy use. This helps them make better design choices while seeing how the solutions change over time (Wilensky, 1999; Lee, 2010).

Still, the results depend a lot on how the problem is set up and approached on the paper and dependent on the designer. But if the goal is not clear or the parameters are too loose, the algorithm might not find good solutions, or it might take too long to reach the optimized point. Also, even if the algorithm finds options that look good in terms of numbers, they may not always work well in real life. That is why designers often need to review the results and make their own judgments. In this way, Genetic Algorithms are best seen as support tools, not final decision-makers.

OBSERVATIONS AND LIMITATIONS

Using Genetic Algorithms in early-stage building design brings both benefits and challenges in the process and in one of the biggest advantages is their flexibility in the approach. Designers can explore many different options without needing to follow one strict path. The algorithm helps them test many combinations of form and performance, which is useful when the design goals are not fixed yet.

Another benefit is that GAs work well with multiple goals in the design phase which could be really helpful for the designer to get to the point they want. For example, in Octopus, designers can try to improve daylight while also reducing energy use and material cost. The tool gives a range of possible solutions, instead of just one best result. This allows designers to compare and choose based on their own priorities. In this way, GAs support creative freedom while still giving feedback on performance.

However, there are also some clear downsides. One of them is the time it takes on the computer. GAs often needs many generations to find useful results, especially when the model is complex or the goals are hard to calculate. This can be a problem if the designer has a deadline or if the software crashes during the process. Also, if the fitness function is not well defined, the results might not make much sense at the end of the simulation which could end in huge time loss.

Another issue is that the algorithm does not understand context. It only sees numbers and tries to improve them. But architecture is not only about performance. It also needs to respond to site, culture, human experience and other parameters which could need to finish the design. Some GA results may look good in terms of performance but might be too strange, hard to build, or just not appropriate. So, the designer still needs to evaluate the output carefully.

CONCLUSION AND OUTLOOK

This paper explores how Genetic Algorithms can support early-stage design in architecture within the tools and approaches to the optimization and their ability to handle complex and open-ended problems makes them a good match for the early phases of design, where many options are still possible. Tools like Galapagos and Octopus have made these algorithms more accessible by allowing visual workflows and clear connections between design goals and performance feedback.

At the same time, using GAs also comes with challenges. It takes time to set up good fitness functions, and the results may not always make sense in real-world projects. Designers still need to guide the process and evaluate the results based on experiences, contexts, and common senses of the design. GAs is not a solution on their own but can act as helpful assistants that offer new ideas and directions.

Looking forward, GAs could become even more useful if combined with other tools like machine learning or real-time feedback systems within Large Language Model's which could help the process get faster a lot and these could help speed up the process or filter out unrealistic options early. Still, human input will always be needed to make final decisions and keep the design grounded. The value of GAs is not just in finding one best solution, but in supporting the designer's thinking with new and unexpected alternatives.

As architectural practice becomes more data-driven and performance-oriented, it is important to critically reflect on how optimization tools are used, mainly for efficiency, but also for creativity and exploration in the field. Genetic Algorithms can open up new ways of thinking about design, but only when they are used with clear goals and thoughtful interpretation. As students and future professionals, understanding both the possibilities and limits of these tools helps us stay in control of the design process rather than being led by it.

REFERENCES

Bentley, P.J. (1999). Evolutionary Design by Computers. Morgan Kaufmann, San Francisco.

Deb, K. (2001). Multi-Objective Optimization using Evolutionary Algorithms. John Wiley & Sons.

Nazari, F. & Yan, W. (2025). Multi-objective optimization of building form using genetic algorithms. Journal of Building Performance Simulation. (forthcoming).

Nazari, F. & Yan, W. (2025). A Case Study on Evaluating Genetic Algorithms for Early Building Design Optimization: Comparison with Random and Grid Searches. Preprint on arXiv.

Fasoulaki, E. (2007). Genetic Algorithms in Architecture: a Necessity or a Trend? MPhil thesis, MIT.

Taleb, H.M., Kayed, M. & Baba, F. (2024). Genetic Algorithm for Optimizing Urban District and Block Morphology to Minimize Solar Radiation Access and Maximize Building Floor Area in the UAE. Buildings, 14(12), 3898.

MDPI BETlab (n.d.). Parametric Design Optimization Using Genetic Algorithms, using Rhino + Grasshopper and Galapagos for daylighting.

Rant, N.N. (2019). Using genetic algorithms in parametric building façade design, Aalborg University thesis.

Su, Z. & Yan, W. (2015). A fast genetic algorithm for solving architectural design optimization problems. Artificial Intelligence for Engineering Design, Analysis and Manufacturing, 29(4), 457–469.

Generative AI for Architectural Floor Plan Design Automation

Lucija Aleksić
TU Wien
lucija.aleksic@tuwien.ac.at

This paper investigates the use of generative artificial intelligence (GenAI) for automating architectural floor plan design, a task traditionally marked by iterative refinement and cognitive complexity. We examine models of different architectures, from GANs/VAEs to the state-of-the-art generative models, including diffusion models and large language model (LLM)-assisted pipelines. The focus of the paper will be their possibilities in synthesizing spatial layouts that satisfy both functional and aesthetic constraints. Two case studies are presented: a diffusion-based framework trained on annotated floor plans that achieves high structural fidelity, and a hybrid LLM-diffusion pipeline that translates natural language prompts into CAD-ready plans. We also address ethical concerns, such as dataset bias, labor impacts, and privacy. The results highlight GenAI's potential to enhance design workflows, underscoring the need for responsible use.
Keywords: *Generative AI, Architectural Design, Floor Plan Generation, Diffusion Models, Large Language Models (LLMs)*

INTRODUCTION

The process of architectural floor plan design is a creative and spatially complex task that usually involves iterative refinement, expert intuition, and domain-specific knowledge. The process balances aesthetic goals with fundamental functional and structural constraints, such as zoning, circulation and daylight access (Yazdandoust, 2024). With the advent of Generative Artificial Intelligence (GenAI) there is a growing interest in automating parts of the early-stage ideation process in architectural design.

Generative AI models are capable of learning and producing new data based on the training data. That means they can be applied to many domains, including the floor plan production. They can satisfy user-defined constraints, and output plans that conform to both functional and aesthetic goals. This automation is not only accelerating workflows but also expanding the creative boundaries of human designers by surfacing unconventional yet viable layouts. This paper explores recent advances in the use of GenAI, specifically diffusion models and large language model (LLM)-assisted pipelines, for automating floor plan generation.

We analyze their architectures, training paradigms, performance outcomes, and ethical implications. In this paper will also be two case studies with actual GenAI application and results.

OVERVIEW OF GENAI MODELS IN ARCHITECTURAL DESIGN

Generative models span over diverse model families, from latent-diffusion networks to modern large-language models, and, irrespective of their architecture each has potential application in architecture domain.

Generative Adversarial Networks (GANs) and Variational Autoencoders (VAEs)

GANs and VAEs were among the earliest models applied to floor plan generation. GAN models learn by training a generator-discriminator pairs so that they better themselves through training. After successful training, only the generator part is taken and used for future generations of (Shelf, 2023). VAE architecture is based on learning how to encode data, in this case structural semantics, into a latent space to enable future sampling from there. Even though both models have an advantage with their simpler architecture and faster training, both models often suffer from limited diversity and mode collapse, and they struggle to capture complex geometric constraints inherent to architectural layouts

Diffusion Models

Diffusion models are generative models that learn how to iteratively transform random noise into structured data, such as images or floor plans, by learning to reverse a gradual noising process through a series of denoising steps guided by a neural network (Yang, Zhilong, Song, 2022). Since they were originally developed for image synthesis they have been easily adapted for floor plan generation. These models iteratively denoise a random latent vector toward a coherent floor plan design, enabling fine-grained control over the generation process.

They have an inherent stability and capacity for high-resolution output. Architectural extensions often include spatially aware attention layers and domain-specific loss functions to improve functional fidelity.

LLM-Assisted Pipelines

Originally designed for natural language processing tasks such as translation and summarization, large language models (LLMs), like GPT models, have slowly overtaken a wide range of knowledge-driven domains, including architecture (Minaee, Mikolov, Nikzad, 2024). Their ability to understand, reason over, and generate structured content from free-form text has made them particularly useful in bridging the gap between human intent and computational design tools.

In architectural applications, LLMs can interpret complex natural language prompts, such as a client's description of spatial needs or lifestyle preferences and translate them into machine-readable formats like JSON. These structured outputs encode semantic design intent, including room types, adjacencies, orientations, and constraints. When combined with downstream diffusion or rendering modules, this creates a powerful hybrid pipeline capable of producing floor plans from high-level textual descriptions.

CASE STUDY 1: PURE DIFFUSION MODELS

The work of this paper introduces an interesting approach to train and apply a diffusion model specifically for floor plan generation (Yeomans, Lipson, 2024). Their model builds upon a modified U-Net backbone, augmented with attention layers that encode spatial and structural constraints such as adjacency, circulation paths, and seismic wall zoning.

The architecture integrates hierarchical features from input noise through a denoising process, guided by architectural priors. Spatial transformer blocks help the model attend to room connectivity and load-bearing wall alignment.

The model is trained on a curated dataset of 50,000 annotated floor plans from the RPLAN repository, ensuring diversity in layout size, room function, and regional standards.

In training a model loss functions tell us how to evaluate model's performance; the higher the loss is, the worse the performance. Defining a good loss function is necessary for quality output. They

have used a compound loss which combines structural similarity index (SSIM) for visual plausibility with mean average precision (mAP) to evaluate the accuracy of functional components, such as correct placement of doors, walls, and structural supports. Basically, it means that the takes into consideration both functional performance of the output plan as well as the aesthetic, while learning how to produce floor plans.
The model achieved 89% accuracy in seismic wall placement compared to expert-designed ground truths. It also outperformed baseline GAN and VAE methods in both aesthetic (SSIM: +12%) and functional metrics (mAP: +17%).

Figure 1
Combined images from (Yeomans, Lipson, 2025) showing the original dataset, first batch, and final output.

The "HouseLLM" framework (Zong, Zhan, Tan, 2024) exemplifies a two-phase text-to-floorplan pipeline that combines the semantic reasoning capabilities of LLMs with the precision of diffusion models.
The first phase is the LLM-Based Semantic planning. That includes the process of generating a representation of a plan based on free text instructions. For example, a prompt such as "Design a 3-bedroom house with a south-facing living room and open kitchen" is processed by GPT-4, which generates a structured JSON layout. This includes room types, sizes, adjacency constraints, and even annotations like "more sunlight exposure for living areas." This phase must ensure all user requirements are met.
In the second phase the authors utilized diffusion-based layout refinement. The JSON output from the previous phase is passed into a modified version of HouseDiffusion, which is a diffusion model, which then converts the semantic representation into a CAD-ready plan. This model integrates functional priors (e.g., bathroom adjacency to plumbing) and architectural realism constraints.
It is also possible to generate layouts from a graph input, as well as only generating in one phase.
Compared to rule-based or procedural models, HouseLLM significantly improved design compliance (e.g., room connectivity and natural lighting orientation). User studies showed higher satisfaction scores in interpreting generated designs (4.6/5 vs. 3.9/5 for procedural baselines).

ETHICAL CONSIDERATIONS

As it usually is, all fast advances in science and technology come with downsides worth considering. As generative AI becomes increasingly integrated into architectural workflows, it brings with it a set of ethical challenges. One of the most pressing issues is bias amplification. GenAI models, as well as all artificial intelligence models, inherit the biases present in their training data, and architectural datasets are no exception. Many commonly used datasets disproportionately represent Western design norms, particularly single-family, detached housing layouts, while underrepresenting communal and informal housing typologies common in other parts of the world. This can lead to generated designs that lack cultural sensitivity or practical relevance in non-Western contexts, ultimately reducing the global applicability of such models (Jang, Suhyung, Roh, 2025).
To mitigate this issue, researchers and practitioners have begun creating more inclusive and representative datasets. Projects such as the GlobalBuildingAtlas provide data of buildings across the globe, including indigenous and low-income housing, which can create rich and diverse training data and helps models generate contextually appropriate designs. Such efforts are crucial to broadening the scope of generative

systems beyond their initial, biased training domains.

Another major concern is the potential displacement of labor within the architectural profession. According to recent industry analyses, up to 23% of conceptual design tasks may be automatable by 2030 due to advances in GenAI technologies. While this could lead to efficiency gains and cost savings, it also threatens roles traditionally held by junior architects and interns, for whom schematic design is a critical learning stage

CONCLUSION

Generative AI is quickly reshaping the field of architectural design, particularly in the domain of automated floor plan generation. By learning spatial hierarchies, functional dependencies, and aesthetic patterns generative models offer a powerful tool for accelerating and enhancing the early stages of architectural ideation. This paper has explored the key advancements in generative technologies, from early GAN and VAE architectures to state-of-the-art diffusion models and LLM-assisted pipelines, highlighting how these approaches are being adapted to meet the unique challenges of architectural layout synthesis.

Through the analysis of two case studies, we demonstrated how diffusion models can generate high-fidelity, functionally accurate plans, and how LLM-guided workflows can bring human users to the plan they had in mind, acting and creating plans based on their instructions. These technologies not only enable efficient exploration of design alternatives but also open up new opportunities for collaboration between human designers and AI systems.

At the same time, the integration of generative AI into architectural practice raises important ethical questions. Bias in training data, the risk of labor displacement, and the potential leakage of sensitive client information underscore the need for responsible deployment. Efforts to curate more diverse datasets and implement privacy-preserving techniques are essential to ensuring that GenAI tools support inclusive, equitable, and secure design processes.

REFERENCES

[1] Generative ai in architectural design: Application, data, and ethics. 2024. White Paper/Report.
[2] Shervin Minaee, Tomas Mikolov, Narjes Nikzad, Meysam Chenaghlu, Richard Socher, Xavier Amatriain and Jianfeng Gao. Large language models: A survey, 2024
[3] Hod Lipson Santiago Yeomans. From abstract noise to architectural form: Designing diffusion models for efficient floor plan generation. 2024.
[4] shelf.io. Gans explained: How generative adversarial networks work, 2023
[5] Ling Yang, Zhilong Zhang, Yang Song, Shenda Hong, Runsheng Xu, Yue Zhao, Wentao Zhang, Bin Cui, and Ming-Hsuan Yang. Diffusion models: A comprehensive survey of methods and applications, 2022.
[6] Ziyang Zong, Zhaohuan Zhan, and Guang Tan. Housellm: Llm-assisted two-phase text-to-floorplan generation, 2024.
[7] Fatemeh Yazdandoust. Optimizing building form for daylight access in dense urban areas in the early stages of design. Journal of Propulsion Technology, 45(3):4145–4157, 2024.

GAN's in Floor and City Block Planning

Insights in how AI can aid and automate planning processes

Paul Scheipner[1]
[1]*Technical University of Vienna*
[1]*paul.scheipner@student.tuwien.ac.at*

Designing both building interiors and urban blocks traditionally relies on manual drafting, iterative reviews, and exhaustive rule-writing-processes that are time-consuming, error-prone, and difficult to scale when complex regulations or local customs apply. In recent years, Generative Adversarial Networks (GANs) have emerged as a powerful, data-driven alternative for automating layout generation at both scales. By learning from thousands of existing examples rather than encoding every building code or zoning ordinance by hand, GANs can capture the implicit expertise embedded in historic floor plans (Goodfellow et al., 2014; Ali et al., 2025) as well as the cultural and legal nuances of city blocks (Fedorova et al., 2021; Xu et al., 2021). We conclude by identifying remaining challenges and promising paths toward fully automated, regulation-aware tools that architects and urban planners can trust and customize.

Keywords: *Generative Adversarial Networks, GAN, Floor & Block Planning, Gen Ai*

INTRODUCTION

Whether one is arranging rooms within a house or plotting parcels in a neighborhood, the core challenge is the same: satisfy intricate rules while crafting layouts that feel natural to end users. Traditional methods rely on architects' manual sketches, rule-based checks ("hallways must be X feet wide," "doors must swing inward"), and iterative revisions-a workflow that becomes exponentially harder as each new code or local convention (e.g., Amsterdam's bike lanes) is added (Smith & Jones, 2018; Fedorova et al., 2021; Xu et al., 2021).

To overcome these bottlenecks, researchers have turned to GANs - which learn implicit layout conventions from data instead of hand crafting every rule (Goodfellow et al., 2014). Early GAN's captured general building outlines but ignored room adjacencies and often produced awkward shapes (Radford et al., 2016; Uzun et al., 2020). Conditional image-to-image models then translated sketches or color-coded masks into walls, doors, and windows, greatly improving realism but sometimes yielding soft, blurry boundaries (Isola et al., 2017; Zheng & Huang, 2018; Sato et al., 2020). More recent graph-based GANs (e.g., House-GAN) accept room-adjacency or parcel-graph inputs, ensuring that generated layouts respect functional relationships and local constraints (Nauata et al., 2020).

GANS'S IN FLOORPLANING

Traditional floor-plan design relies on architects' manual drafting and iterative revisions-a process that takes a long time, is prone to mistakes, and becomes especially difficult when complicated building codes and regulations come into play. Early attempts to automate this work used sets of explicit rules-say, "hallways must be at least X feet wide" or "doors must swing in a certain way"-but adding each new requirement made the system exponentially more complex, and it quickly became impractical for real-world projects. In response, researchers have turned to Generative Adversarial Networks, or GANs, which learn from examples instead of relying on hand-written rules. By studying large collections of existing floor plans, GANs can pick up on the unwritten know-how of expert designers and produce layouts that feel familiar without explicit programming (Goodfellow et al., 2014).

Automating floor-plan creation still has its unique challenges. In the past, rule- or agent-based systems were popular, but each additional regulation-such as those for hospitals with strict safety rules-multiplied the system's complexity by a large factor, making it hard to maintain or extend (Smith & Jones, 2018). By contrast, a data-driven GAN model can absorb these complexities simply by including more examples during training. Real-world plans embed subtle conventions-like which rooms tend to sit next to each other, common paths people walk, and minimum room sizes-so a GAN trained on thousands of plans can learn these patterns naturally (Ali et al., 2025).

The first GANs applied to floor plans were based on Deep Convolutional GANs (DCGANs). These models take a random "seed" and try to turn it into a plausible building outline. While they could capture the general look of buildings in the training set, they often produced rooms in odd shapes and ignored important relationships like how bedrooms connect to bathrooms (Radford et al., 2016; Uzun et al., 2020). In other words, DCGANs showed promise but couldn't yet replace careful human planning.

Next came conditional image-to-image GANs like Pix2Pix, which treat floor-plan creation as a translation task: you give the model a simple sketch or color-coded zones (for example, blue for bathrooms, yellow for kitchens) and it fills in the walls, doors, and windows. This approach produces much more detailed and usable layouts (Isola et al., 2017). However, because these models typically use basic "pixel-by-pixel" comparison rules during training, their outputs can look a bit soft around the edges-walls may appear blurry, and fine details like window frames can be smudged (Zheng & Huang, 2018; Sato et al., 2020).

To make sure rooms appear in the right places relative to one another, researchers developed graph-based GANs such as House-GAN. Instead of pixels, these models take a simplified "bubble diagram" showing which rooms should connect. They then produce floor plans that respect those connections, and studies have shown people find these layouts more realistic and varied compared to simpler GANs (Nauata et al., 2020). The catch is that the results are still flat images, which makes it hard to import them directly into architectural software that prefers clean lines and editable objects.

Another practical effort, NVIDIA's ArchiGAN stack, breaks the problem into three steps: first it draws the overall building shape based on a property map, then it divides that shape into rooms with doors and windows, and finally it suggests furniture layouts (Chaillou, 2019). This step-by-step pipeline lets architects review and tweak each stage, combining fast AI suggestions with human expertise. While the demos look impressive, there isn't yet hard data on how often

architects accept these suggestions or how much time they save.
More recently, people have looked closely at the "rules" used to teach GANs. Instead of simple pixel-difference measures, they've tried a more forgiving formula called the Huber loss, which balances emphasis between small and large errors. On a large test set of sample plans, GANs trained with Huber loss produced crisper walls, cleaner room boundaries, and more reliable results than those trained with conventional methods (Ali et al., 2025).

Together, these developments show a clear path forward: moving from rigid, rule-based tools to flexible, data-driven methods that learn directly from real plans, enforce room relationships, and offer human-in-the-loop control. The next big step will be turning these flat images into editable, vector-based layouts that plug straight into architects' software. Promising ideas include hybrid models that blend graph reasoning with image translation, new generative techniques that output vector graphics directly, and interactive interfaces where designers can guide the AI in real time. By bringing these advances together, GANs in floorplanning may soon provide architects with fully automated, regulation-aware tools they can trust and customize.

GAN'S IN CITY BLOCK PLANNING

Even though planning a city block operates at a whole different meta level than planing individual floor layouts, the underlying challenge of adhering to (legal) regulations and meeting inhabitants' needs is quite similar (Goodfellow et al., 2014; Fedorova et al., 2021). Consequently, just as floorplan automation shifted from rule-sets/agents to data-driven methods, researchers have begun exploring GAN's to learn city-block patterns directly from existing urban patterns and data. This – at least for a large part - sidesteps the combinatorial explosion of hand-coded constraints and (human)effort(Goodfellow et al., 2014).

Moreover, city blocks introduce an extra dimension of complexity: local culture. For instance, designing a block in Amsterdam might need to consider accommodating bicycle traffic, while city block in a US city might be more considerate to cars (Fedorova et al., 2021; Xu et al., 2021). As a result, any automated system must not only respect building-height limits, setback rules, and land-use ratios, but also must be able to consider these implicit 'rules'.

Early work demonstrated that even a relatively simple GAN can internalize a city's morphological rules. In one study, Fedorova et al. (2021) trained a GAN on OpenStreetMap footprints from Milan, Tallinn, and Bengaluru, and showed that, without any explicit programming, the model learned block-shape distributions, street-alignment tendencies, and plot-size variations unique to each city. Consequently, the generated block layouts felt contextually appropriate, whether in the narrow lanes of Tallinn's old town or the winding streets of Bengaluru (Fedorova et al., 2021).

Building on this, Xu et al. (2021) introduced BlockPlanner, which represents each parcel and its neighboring lots as nodes in a two-tier graph. By combining a lightweight graph-VAE with GAN components, BlockPlanner produces full 3D block proposals that honor both plot boundaries and adjacency relationships, while also suggesting building-height variations that match real zoning codes. In comparative evaluations, this graph approach resulted in a more realistic land-use distributions and smoother transitions between neighboring plots than pixel-only methods (Xu et al., 2021).

In yet a completely different approach, Text2City (Qin et al., 2024) reframes block planning as a text-conditioned diffusion problem. By feeding in descriptions like "mixed-use waterfront district with pedestrian promenades," alongside surrounding map context, the model regenerates coherent road networks and building footprints that blend seamlessly with existing urban elements. This one-stage approach not only accelerates block prototyping but also allows non-technical people to perform urban design through natural language. As most city planners are probably not programmers, this novel approach makes a lot of sense for ease of use and acessablity (Qin et al., 2024).

Meanwhile, Gan et al. (2024) proposed UDGAN, which marries style-learning GANs with multi-objective optimization. By capturing a planner's aesthetic preferences - such as courtyard proportions or facades - UDGAN can generate multiple block schemes and then rank them according to predefines metrics. These intuitive metrices and KPI's are something like solar access and walkability. As a result, designers gain a palette of viable options that align with both regulatory requirements and personal design intentions (Gan et al., 2024).

Finally, UrbanGenoGAN (Cheng et al., 2023) demonstrates how GANs can integrate with evolutionary algorithms and GIS-based spatial analysis. A GAN proposes initial block layouts, and secondly a genetic algorithm refines them for connectivity and green-space distribution, and additionally, GIS tools verify flood-risk and transit accessibility. In case-study comparisons, UrbanGenoGAN outperformed conventional manual and optimization-only workflows on both quantitative indicators and planner satisfaction scores (Cheng et al., 2023).

Together, these advances trace a clear progression: from early GANs that learned generic block shapes to sophisticated, context-aware frameworks that incorporate graphs, text prompts, style preferences, and spatial analytics. Although challenges remain-particularly in exporting raster outputs to editable vector formats-promising directions include hybrid graph-pixel architectures, diffusion models with native SVG outputs, and interactive interfaces that let planners iteratively steer the generative process.

CONCLUSION

The unrelated evolution of GAN-based methods for floor-plan and city-block generation has a similar trajectory: from early, unconstrained models to conditional, graph-aware, text-guided, and multi-stage systems that embed regulations, culture, and human expertise directly into the learning process. Despite already impressive gains in realism, coherence, and user control, a massive gap remains in converting rasterized outputs into editable formats that integrate seamlessly with architects' CAD tools or urban planners' GIS platforms.

Promisingly moving forward, the synthesis of graph and pixel representations, the rise of vector-friendly diffusion models, and the design of interactive interfaces promise to bridge this very problem. By uniting these advances, the next generation of GANs might become truly automated, regulation - and culture-aware layout tools that enable designers at all scales to create safe, efficient, and contextually resonant environments without requiring the operator to be a computer scientist. In the ideal outcome, these GANs and systems will enable architects and planners-removing redundant and repetitive work-rather than replacing them themselves.

Wave Function Collapse for Architectural Applications

An analysis of the algorithm, its potential and existing integrations

Laura Müllner
Technische Universität Wien
e11701293@student.tuwien.ac.at

The wave function collapse algorithm, which has gained popularity particularly in independent video game development, its conception and concrete implementations for the creation of example-based procedural content generation are analyzed. Its ability to generate comparatively large, complex and locally similar output models from minimal input is of interest in the context of generative architecture, where machine learning approaches successful in other domains have had limited success due to a lack of publicly available training data. Current research on architectural applications is investigated and future potential discussed.

Keywords: *Wave Function Collapse, Model Synthesis, Procedural Architectural Generation, Generative Architecture, Constraint Satisfaction Problem.*

INTRODUCTION

As digital models have gained significant relevance in the entertainment industry, speeding up the highly skilled and time consuming task of creating these models is in particularly high demand for use in Computer Generated Imagery and video games. Especially the latter field has employed different forms of Procedural Content Generation (PCG) since the late 1970s, as it additionally benefits from reduced game file sizes.

Similarly, the establishing of rulesets or shape grammars with the goal of floorplans, models and designs has been a target of research in the architecture field beginning with the emergence of computational methods (Stiny and Mitchell, 1978; Weber, Mueller and Reinhart, 2022). Automatically iterating and optimizing the laborious task of creating suitable room layouts however necessitates the combination of a multitude of different factors, considering aesthetic, functional, economic, ergonomic, physiologic and structural aspects. As a result, such methods have still not become mainstream in architectural practice, although current developments with Deep Learning Models might soon change this (Weber, Mueller and Reinhart, 2022).

The so-called „Wave Function Collapse Algorithm" (WFC) is a type of constraint-based, generative algorithm that can be used to create complex procedural, three-dimensional models based on comparatively simple input objects. It has found rapid adaptation in the video game industry for the purpose of 2D and 3D content generation, as it allows for the automatic creation of entire levels - such as in the shape of room sequences, cities or landscapes - from a limited set of input tiles. The similarity of the output model to its input and the often inherently spatial or even specifically architectural nature of these implementations invites the application of the algorithm in real-life architectural context.

Existing research on such integrations in the design process as well as the algorithm's functionality are analyzed within this paper.

DEVELOPMENT

With the research on computer graphics as a result of the progress of digital technologies in the 1980s and -90s various methods for procedural generation of 2D and 3D assets were developed. Based on work for 2D texture synthesis (Efros and Leung, 1999), Paul Merrell created an algorithm for example-based 3D "Model Synthesis" (Merrell, 2007). This discrete approach extended upon local region growing methods in two dimensions by subdividing a given input model into so-called "model pieces". These can subsequently be used as three-dimensional tiles out of which a newly generated model is assembled, much like pixels in texture synthesis.

To ensure a logically consistent output model, the input is automatically analyzed for existing model piece adjacencies. They are set as constraints for the procedural generation, such that only tiles adjacent in the input data may occur in adjacent positions within the output. During the generative process, these adjacencies restrict the domain of possible choices for each grid space in the output model based on pieces already placed, converging the solution space until a consistent model has been generated (Paul C. Merrell, 2009).

Based on Discrete Model Synthesis, Maxim Gumin published several implementations of two-dimensional texture synthesis algorithms on his GitHub page (Gumin, 2016), which rapidly found an audience among independent game developers. Apart from the "Simple Tiled Model", which uses Model Synthesis' direct neighbour adjacency constraints, he extended upon the constraint generation in a second "Overlapping Model", which considers a wider matrix of $N \times N$ neighbouring tiles and is primarily used in 2D applications.

As the iterative reduction of possible values until convergence in the generated output matrix roughly resembles the concept of wave function collapse due to observation in quantum mechanics, Gumin coopted the term for the procedural algorithm (Karth and Smith, 2022). Despite his algorithm resembling Merrell's in all its core functionality, as the repository's popularity led to the algorithm being known as "Wave Function Collapse" it will subsequently be referred to under that name.

The generative approach gained widespread popularity following the release of several indie games utilizing it for its level design, including Oskar Stålbergs Bad North (2018) and Townscaper (2021) (Stålberg, 2021).

FUNCTIONALITY

While specifics of the algorithm may vary from implementation to implementation, core generative principles largely resemble those utilized in the original Discrete Model Synthesis approach.

Adjacencies between model pieces function as the constraints based on which the outcome is generated. To determine these, the input model must be tiled, that is, consist of a discrete set of model pieces or pixels that can be uniquely identified and reassembled in the output. During the generation, the algorithm iterates through two phases, the observation and the propagation stage, until a consistent solution has been produced. Depending on the implementation and application of the algorithm, additional Pre- and Post-Constraints or additional processes may be added to control possible outcomes (Karth and Smith, 2022).

Adjacency

In the initial step of the Model Synthesis algorithm valid adjacencies are extracted from the input data. For k unique tiles in the input model E, binary transition matrices for each dimension of size $k \times k$ are initialized with every existing adjacency between two tiles (Paul C. Merrell, 2009). In the case of an overlapping WFC approach, a secondary tile set of combined tile patterns is created as an intermediary step, for which the adjacency matrices are then calculated (Karth and Smith, 2022).

Additionally, alternative or even manual assignment of adjacency rules to a discrete set of tiles can be used as a more structured approach when an example model containing all desired combinations might be unfeasible (Tigas and Hosmer, 2021; Jiang and Wang, 2023; Zhang *et al.*, 2025). Automated approaches can be based on matching section curves or profiles (Khokhlov, Koh and Huang, 2019; *Konsoll 2018: Oskar Stålberg - Developing The Bad North Look*, 2019).

As the variation of neighbouring tiles and patterns in the input model directly determines valid outputs, WFC manages to create models with strong local similarity to any given input. In cases where only direct adjacencies are considered, a more global resemblance can further be enforced during the propagation stage (Merrell, 2007).

Observation and Propagation

For the generation of the output, a matrix of selected dimensions and size is initialized with every slot containing all possible tile identifiers. An initial tile may be placed manually or at random, removing all but the set tiles' identifiers from its location, and reducing directly neighbouring slots based on the adjacency constraints. This decrease in valid identifiers propagates through the matrix, recursively removing all labels whose required adjacencies can no longer be met by the ones left in the reduced domain. The amount of tile identifiers remaining as possible values for a given space is referred to as "entropy".

Once the propagation stage is complete, the matrix is observed to determine the slot with lowest entropy to be selected as the next one to collapse. A remaining label is chosen at random and the constraints are again propagated through the matrix (Merrell, 2007; Karth and Smith, 2022).

Backtracking and Conflict Resolution

This lowest-entropy-approach reduces possible conflicts during the generation, which may occur when all tile identifiers are removed from a slot within the matrix (Karth and Smith, 2022). The probability of a model becoming inconsistent depends on the dimensions of the output model, variation in the tile set and given constraints and is generally considered an NP-complete problem.

To avoid this, the generation process may be intermittently saved and restarted from a previous state. Additionally, the output matrix can be initialized complete with empty tiles, which is a known consistent solution, and regenerated in smaller, overlapping sections, which ensures short computational times before consistency can be determined (Merrell, 2007). Other approaches involve entirely restarting (Karth and Smith, 2022) or accepting inconsistencies as part of the intended outcome (Koh, 2022).

Pre- and Post Constraints

Added constraints applied either before the model generation or as a post-processing step can improve control over the outcome of general WFC.

A common approach involves pre-seeding the result matrix by either reducing the domain or entirely collapsing slots to ensure specific tile placements. This may be used to ensure floor or foundation tiles are positioned at the ground plane (Paul C. Merrell, 2009), ensure the correct placement of vital tiles (Boris, 2020) or specify areas (Zhao *et al.*, 2023). Provided an accessible interface, this could allow for an interactive and collaborative design process, in which designer or user input is supplemented by auto-propagation or partial collapsing via the algorithm.

Adjusting the individual weights or probabilities of tiles can ensure certain distributions or rarity of specific model elements. By correlating the probability depending on the location within the matrix, the overall shape of the final model can be constrained (Merrell, 2007).

Path or connectivity constraints can be used to ensure accessibility of all rooms or spaces within the model, which force a subset of tiles to create a single connected component (Boris, 2020).

Post-processing steps and evaluation procedures can refine outcomes, help select suitable results and, particularly in combination with machine learning, iteratively improve the generation workflow itself. Koh (2022) uses a "structural connectivity" algorithm to identify and remove floating model pieces. A simplified tile set used during the generative process may be enhanced based on its final position and context (Karth and Smith, 2022), which can be used to place borders, gradients or perhaps walls at intersections of differently tiled areas.

ADAPTATION FOR ARCHITECTURE

WFC combines both rule-based as well as data-driven approaches commonly utilized for computational solutions to Space Allocation Problems, which include both floorplan generation at an architectural level as well as urban planning at a larger scale (Zhang *et al.*, 2025). Rule or grammar based approaches to generating architecture, while well represented in research (Stiny and Mitchell, 1978; Wonka *et al.*, 2003; Müller *et al.*, 2006) tend to restrict the generated model based on the original developers intent for it. This makes incorporation into a planning workflow difficult, as a design user generally has neither knowledge nor control over the generative process or the rules governing the properties of the model.

Due to the WFC algorithm's ability to work based off a set of conventional 3D models which are already a common part of digital design workflows, it presents a more intuitive approach to procedural generation that can take advantage of existing knowledge and skills. Designers have full control over the input tiles, which can be customized and evaluated on a visual basis with familiar tools such as Rhino and Grasshopper.

Monoceros

Developed by Ján Pernecký and Ján Tóth at Subdigital, Monoceros is a Grasshopper addon that implements the tools for tile definition, adjacency constraints as well as constraint solving according to Gumin's WFC algorithm (*Monoceros*, no date).

It innovates upon the Simple Tiled Model by allowing for the creation of Modules, which are fixed combinations of multiple tiles. These can be used to define specific rooms or connected structures, increasing the designer's control over the final model in a trade-off with associated efforts for setup as well as a reduction in variable outcomes.

As the underlying program Rhinoceros 3D is commonly used in architectural practice and provides a visual programming interface designers are more likely to be familiar with, this plugin represents a significant step in making WFC as a tool accessible to practicing designers.

It further increases the concepts visibility among architectural professionals, whose valuable insights on the logic behind procedural and qualitative parameters have so far been restricted by knowledge on advanced programming acting as a barrier of entry. Its release also enables research to focus on the extension of the WFC functionality beyond the core algorithm without the additional effort of having to develop an implementation suitable for common architectural use.

Module-based WFC

Zhang et al. (2025) utilize a combination of WFC, quantitative analysis and multi-criteria decision-making to optimize rural land usage as well as two-dimensional residential layout generation.

Based on floorplans collected from rural Chinese areas, average room types, sizes, dimensions, distribution and connectivity are extracted. Using Monoceros a room module library covering these parameters is automatically generated from the base tiles, which can be filtered based on size, type or aspect ratio. Within a site boundary determined via generative solver a floorplan layout can then be generated, selecting a subset of suitable room modules based on the statistical evaluation of real-world examples.

The pre-constraining of available tilings and modules reduces computational requirements, increases realism of results and simultaneously provides a method of additional control to the user. Generated layouts are automatically evaluated within Grasshopper by layout rationality, circulation simplicity and room layout rationality. This multi-step approach demonstrates a successful integration of WFC in a data-driven, generative architectural workflow (Zhang *et al.*, 2025).

Voxel-Based Model Generation

As an extension of the overlapping model of the 2D WFC algorithm, Voxel Synthesis for Architecture extracts patterns of given cubic dimensions from a given voxel input model alongside their adjacency relations. From the library of patterns an output model is generated in the familiar WFC-fashion, after which post-processing algorithms are applied in order to ensure the structural connectivity of voxels. Further consideration is given to the relation of the model to the human scale, prioritisation of generation from the ground up as an abstraction of gravity for the generative process as well as a parameter allowing continuation despite an adjacency contradiction. (Koh, 2022)

Tile Set Analysis

Zhao et al. (2023) investigated the practical application of so called "discrete design" enabled via WFC and Monoceros by establishing necessary criteria for utilized tile sets for architectural layout generation. Using a two-story single-family home as a design study, they determined four properties for tiles that need to be considered during a generative design approach: slot size, tile topology, tile form and associated rules. Tiles encompassing around a quarter to a ninth of a full room seemed to create the most diverse layouts while keeping rulesets manageable. The tile topology is used to categorize tiles based on the presence and arrangements of walls along the outer edges as well as within their interior, categorically resembling the capital letters OTILUX (see Zhao et al., 2023). Particularly for elements comprising the building envelope such as façade and roof tiles, they suggest the use of different tile forms, representing aesthetic variations of tiles of equal function, as this helps avoid a monotonous or grid like appearance of the final design.

The differentiation between the buildings interior and exterior via the creation of a separate tile set for the building envelope improved the integrity and realism of design outcomes. The output space is predefined by setting up tile slot areas for interior, two separate façade sets as well as the roof area.

They determined four rules as essential for the creation of proper designs within their generative framework: The manual placement of staircases ahead of the automated generation process, the setup of predefined rule templates, the conditional placement of windows to ensure daylight as well as additional rules to avoid illogical door placements.

Evaluation of the generated designs determines room connectivity, convexity, daylight, area as well as corner count and statistically evaluates the probability of tiles and room templates being used as well as a cluster analysis of design outcome similarity. This allows designers to determine necessary changes to tile and rule sets.

Furthermore, two approaches are proposed to integrate functional programs within the generated structure. As the inclusion of furniture enabling architectural programs within the tiles would massively increase tile count and associated computational and modelling efforts, a neural network is used to predict possible room labels with limited results.

Integration of Reinforcement Learning

Mintrone and Erioli (2021) utilize an Artificial Neural Network (ANN) trained via Proximity Policy Optimization (Schulman *et al.*, 2017) to improve upon the WFC solver. Instead of randomly selecting one of the values remaining for the lowest entropy node, the selection choice is learned from and performed by the reinforcement learning algorithm. The reward function is based on a set of six spatial descriptors: Density, spatial distribution, orientation as well as structural, spatial and planar connectivity. While their approach effectively enforces their parameters, the resulting 3D assemblages, made from a comparatively extremely limited set of 6 planar elements and their rotations, are highly abstract. This complicates both visual evaluation by a human designer as well as refined analysis on anything but their predefined spatial descriptors.

A similar approach of utilizing automated 3D WFC in combination with Reinforcement Learning was realized by Tigas and Hosmer (2021). The algorithm is improved with the specific goal of maximizing designer-set objectives, thus reestablishing control over the generation process. The usually random selection of a fitting tile for the node with lowest entropy is again replaced by a policy weighing remaining tiles according to their learned probability to positively influence the outcome. One reward signal used was the maximum displacement of the model following a structural simulation. Using Proximal Policy Optimization the value function and policy for the upcoming round are updated (Tigas and Hosmer, 2021). This approach was utilized in three separate design projects at the Bartlett School of Architecture.

CONCLUSION

Despite its limitations, the Wave Function Collapse algorithm shows promise for its use in architecture. Due to the potential for varied procedural outcomes based entirely on a limited set of designed – and as such controllable – input models it can be utilized where more generalized generative approaches may struggle due to limited training data. By extending the generative process with pre- and post-constraints or additional machine learning steps the inherent functionality may be significantly enhanced for architectural workflows.

Improvements to the available tools could further reduce time and programming knowledge required for such implementations. Versioning for tiles that only differ in orientation or room program could decrease manual effort for rule setup and tile count. Automated support for manual constraint definition and coordination between tiles would help reduce unintended constraints (Zhao *et al.*, 2023).

Many potential extensions of the WFC algorithm have not yet been investigated for possible uses in architectural PCG.

An iterative WFC process, in which the model is assembled in stages out of tiles reducing in scale, from the general massing, to different functional areas, individual rooms down to furniture within those rooms may help overcome the discrepancies between attention to structural integrity and anthropomorphic quality that were apparent in approaches focused on either one. As this would allow for a combination of tiles of different sizes it would reduce the amount of tile variations needed for a single scale.

Furthermore there is very limited research on an alternative Continuous Model Synthesis Approach (Merrell and Manocha, 2008). While implementation is more complex and less intuitive, the variable sizing and constraining angles could be seen as the vectorized approach to the "rasterization" of Discrete Model Synthesis relying on a regular grid. Related research is directly linked to graph-grammars and example-based inverse procedural modelling (Bokeloh, Wand and Seidel, 2010; Merrell, 2023) and may due to its inherent parametricity offer opportunities for a more direct integration with conventional BIM software (Demir, Aliaga and Benes, 2016; Lu et al., 2024).

REFERENCES

Bokeloh, M., Wand, M. and Seidel, H.-P. (2010) 'A connection between partial symmetry and inverse procedural modeling', *ACM Transactions on Graphics*, 29(4), pp. 1–10. Available at: https://doi.org/10.1145/1778765.1778841.

Boris (2020) 'Wave Function Collapse tips and tricks', *BorisTheBrave.Com*, 8 February. Available at: https://www.boristhebrave.com/2020/02/08/wave-function-collapse-tips-and-tricks/ (Accessed: 3 June 2025).

Demir, I., Aliaga, D.G. and Benes, B. (2016) 'Proceduralization for Editing 3D Architectural Models', in *2016 Fourth International Conference on 3D Vision (3DV). 2016 Fourth International Conference on 3D Vision (3DV)*, Stanford, CA: IEEE, pp. 194–202. Available at: https://doi.org/10.1109/3DV.2016.28.

Efros, A.A. and Leung, T.K. (1999) 'Texture synthesis by non-parametric sampling', in *Proceedings of the Seventh IEEE International Conference on Computer Vision. Proceedings of the Seventh IEEE International Conference on Computer Vision*, Kerkyra, Greece: IEEE, pp. 1033–1038 vol.2. Available at: https://doi.org/10.1109/ICCV.1999.790383.

Gumin, M. (2016) 'Wave Function Collapse Algorithm'. Available at: https://github.com/mxgmn/WaveFunctionCollapse (Accessed: 2 June 2025).

Jiang, W. and Wang, J. (2023) 'Stylized Space Synthesis: Exploring the Stylized Generative Design Method of Architecture Based on Wave Function Collapse Algorithm', in. *CAADRIA 2023: Human-Centric*, Ahmedabad, pp. 311–320. Available at: https://doi.org/10.52842/conf.caadria.2023.1.311.

Karth, I. and Smith, A.M. (2022) 'WaveFunctionCollapse: Content Generation via Constraint Solving and Machine Learning', *IEEE Transactions on Games*, 14(3), pp. 364–376. Available at: https://doi.org/10.1109/TG.2021.3076368.

Khokhlov, M., Koh, I. and Huang, J. (2019) 'Voxel Synthesis for Generative Design', in J.S. Gero (ed.) *Design Computing and Cognition '18.* Cham: Springer International Publishing, pp. 227–244. Available at: https://doi.org/10.1007/978-3-030-05363-5_13.

Koh, I. (2022) 'Voxel Synthesis for Architectural Design', in J.S. Gero (ed.) *Design Computing and Cognition'20.* Cham: Springer International Publishing, pp. 297–316. Available at: https://doi.org/10.1007/978-3-030-90625-2_17.

Konsoll 2018: Oskar Stålberg - Developing The Bad North Look (2019). Available at: https://www.youtube.com/watch?v=6JcFbivo8dQ (Accessed: 18 May 2025).

Lu, O. *et al.* (2024) 'Clicking is All You Need: Implementing Wave Function Collapse in Early-Stage Design for Manufacturing and Assembly Projects', in. *CAADRIA 2024: Accelerated Design*, Singapore, pp. 303–312. Available at: https://doi.org/10.52842/conf.caadria.2024.1.303.

Merrell, P. (2007) 'Example-based model synthesis', in *Proceedings of the 2007 symposium on Interactive 3D graphics and*

games. *I3D07: Symposium on Interactive 3D Graphics and Games 2007*, Seattle Washington: ACM, pp. 105–112. Available at: https://doi.org/10.1145/1230100.1230119.
Merrell, P. (2023) 'Example-Based Procedural Modeling Using Graph Grammars', *ACM Transactions on Graphics*, 42(4), pp. 1–16. Available at: https://doi.org/10.1145/3592119.
Merrell, P. and Manocha, D. (2008) 'Continuous model synthesis', *ACM Transactions on Graphics*, 27(5), pp. 1–7. Available at: https://doi.org/10.1145/1409060.1409111.
Merrell, P. and Manocha, D. (2010) 'Example-based curve synthesis', *Computers & Graphics*, 34(4), pp. 304–311. Available at: https://doi.org/10.1016/j.cag.2010.05.006.
Mintrone, A. and Erioli, A. (2021) 'Training Spaces - Fostering machine sensibility for spatial assemblages through wave function collapse and reinforcement learning', in. *eCAADe 2021: Towards a New, Configurable Architecture*, Novi Sad, Serbia, pp. 17–26. Available at: https://doi.org/10.52842/conf.ecaade.2021.1.017.
Monoceros (no date). Available at: https://monoceros.sub.digital (Accessed: 3 June 2025).
Müller, P. *et al.* (2006) 'Procedural modeling of buildings', *ACM Transactions on Graphics*, 25(3), pp. 614–623. Available at: https://doi.org/10.1145/1141911.1141931.
Paul C. Merrell (2009) *Model Synthesis*. University of North Carolina at Chapel Hill.
Schulman, J. *et al.* (2017) 'Proximal Policy Optimization Algorithms'. arXiv. Available at: https://doi.org/10.48550/arXiv.1707.06347.
Stålberg, O. (2021) *SGC21- Beyond Townscapers, YouTube*. Available at: https://www.youtube.com/watch?v=Uxeo9c-PX-w&t=1091s (Accessed: 18 May 2025).
Stiny, G. and Mitchell, W.J. (1978) 'The Palladian grammar', *Environment and Planning B: Planning and Design*, 5(1), pp. 5–18. Available at: https://doi.org/10.1068/b050005.
Tigas, P. and Hosmer, T. (2021) 'Spatial Assembly: Generative Architecture With Reinforcement Learning, Self Play and Tree Search'. arXiv. Available at: https://doi.org/10.48550/arXiv.2101.07579.
Weber, R.E., Mueller, C. and Reinhart, C. (2022) 'Automated floorplan generation in architectural design: A review of methods and applications', *Automation in Construction*, 140, p. 104385. Available at: https://doi.org/10.1016/j.autcon.2022.104385.
Wonka, P. *et al.* (2003) 'Instant architecture', *ACM Transactions on Graphics*, 22(3), pp. 669–677. Available at: https://doi.org/10.1145/882262.882324.
Zhang, Yin *et al.* (2025) 'Rural residential layout generation and evaluation: An approach based on the Wave Function Collapse algorithm', *International Journal of Architectural Computing*, p. 14780771251326773. Available at: https://doi.org/10.1177/14780771251326773.
Zhao, H. *et al.* (2023) 'A Framework for the Design and Evaluation of Architectural Tilesets', in. *eCAADe 2023: Digital Design Reconsidered*, Graz, Austria, pp. 491–500. Available at: https://doi.org/10.52842/conf.ecaade.2023.2.491.

Design Optimization of Mycelium-Based Acoustic Panels Using Algorithmic Planning

An Algorithmic Framework for Acoustic Optimization Using Mycelium

Sama Feiziazar
Technical University of Vienna
e12333478@student.tuwien.ac.at

Mycelium-based composites emerge as an innovative biomaterial for architectural acoustics, offering a sustainable alternative to traditional synthetic acoustic panels. This paper explores how algorithmic planning and parametric design techniques – implemented in Grasshopper – can optimize the design of mycelium-based acoustic panels for improved sound absorption and efficient material usage. A theoretical workflow is presented, encompassing shape generation, acoustic simulation, and performance evaluation in an iterative loop. In absence of experimental case studies, a conceptual example is discussed where panel geometries are digitally optimized to balance acoustic performance with minimal material volume. The results indicate that computational design methods can identify panel configurations that enhance acoustic absorption while using less mycelium material, underscoring the potential of combining novel biomaterials with performance-driven design. This work contributes a methodology for sustainable acoustic panel design and sets the stage for future physical prototyping and validation.

Keywords: *Mycelium-based composites, Acoustic panels, Parametric design, Optimization, Sustainable materials*

INTRODUCTION

Acoustic comfort is a key consideration in architectural design, often addressed by installing sound-absorptive panels to reduce echoes and noise. Conventional acoustic panels are commonly made from mineral fibers or petrochemical foams, materials which pose environmental and health concerns (e.g. non-biodegradability and irritant fibers)(Arenas,2010) In the pursuit of sustainable design, researchers have been investigating renewable and biodegradable materials for acoustics (Arenas & Sakagami, 2020) One promising candidate is mycelium-based composite, a lightweight bio-material grown from fungal mycelium and agricultural waste. Mycelium composites can replace synthetic acoustic materials while offering a biodegradable lifecycle(Waler & Gürsoy,2022), aligning with goals of reducing pollution and landfill waste. However, optimizing the acoustic performance of mycelium panels requires careful design of their form and deployment in space, due to the strong dependence of sound absorption on geometry and surface configuration.

Advances in **algorithmic design** provide new opportunities to tackle this optimization problem. Parametric design tools such as Grasshopper allow architects to generate complex panel geometries and distributions, and to integrate acoustic simulations directly into the design process(Vlaun & Waart,2016). By coupling parametric models with acoustic analysis and optimization algorithms, it becomes feasible to **automatically refine panel designs** to meet acoustic targets while minimizing material usage. This paper presents a framework for design optimization of mycelium-based acoustic panels using algorithmic planning. The aim is to leverage parametric modeling, simulation, and evolutionary solvers to identify panel configurations that maximize sound absorption performance for minimal material input. This paper will provide a theoretical overview of mycelium as a sustainable acoustic material and detail a computational workflow for panel shape and layout optimization. A conceptual simulation example is discussed in instead of a case study, given the lack of experimental data. Through this exploration, we demonstrate how combining an innovative biomaterial with performance-driven design methods can contribute to sustainable and effective acoustic solutions in architecture.

Mycelium as an Acoustic Material

Mycelium-based composites are formed by growing fungal mycelium through organic waste like straw or husks. As the fungi grow, they bind the substrate into a lightweight, porous matrix that is dried to form a solid panel. These panels are biodegradable, low-energy to produce, and fully compostable, making them a sustainable alternative to synthetic acoustic materials. (Walter& Gürsoy,2022).

In fact, commercial products have begun to appear – for example, the company *Mogu* produces interior acoustic panels made of mycelium composite, with published sound absorption ratings for several designs (Mogu, n.d.) Crucially, mycelium-based panels exhibit favorable acoustic absorption characteristics. As a porous, fibrous material, mycelium naturally absorbs sound across mid to high frequencies similarly to other fibrous absorbers. Studies have reported that mycelium composites can achieve high sound absorption coefficients; for instance, boards grown with *Pleurotus ostreatus* fungus reached an absorption coefficient of approximately 0.87–0.99 at 1000 Hz in impedance tube tests. (Gezer & Kuştaş, 2024)

Notably, mycelium foam has shown **strong low-frequency absorption** (below ~1500 Hz), even outperforming traditional materials like cork or mineral fiber panels in certain metrics (Pelletier *et al.*, 2019)demonstrated that mycelium-based acoustic boards can match the performance of petrochemical foam panels around the important 1000 Hz .These findings suggest mycelium panels can effectively reduce noise and reverberation in spaces, making them viable for applications such as cinemas, studios or classrooms. Moreover, mycelium offers benefits of being fire-resistant and non-toxic, further supporting its use in interiors.

Algorithmic Design Optimization

Parametric and algorithmic design techniques enable the precise tuning of acoustic panel geometry. Within platforms like Grasshopper, designers can vary shape parameters—thickness, curvature, patterning—and evaluate acoustic behavior using simulation tools like Pachyderm.

Integrating optimization algorithms (e.g., Galapagos) allows automated refinement of panel configurations to reduce reverberation time or improve absorption. Previous studies have successfully used such workflows to optimize acoustic treatments in interior spaces. Multi-objective tools like Octopus support balancing conflicting goals, such as maximizing absorption while minimizing material usage.

Methodology

This study proposes a conceptual workflow for optimizing the design of mycelium-based acoustic panels using parametric and algorithmic design tools. The methodology is structured into four main stages: parametric design of panel geometry, mycelium material acoustic performance, acoustic simulation in design, and evaluation/optimization strategies.

Parametric Design Framework for Acoustic Panels:
Parametric design platforms like Rhino/Grasshopper allow for rapid iteration of acoustic panel geometries by adjusting key parameters such as thickness, surface modulation, and spacing. Studies by Wiggins (2023) and Peters et al. (2021) emphasize the importance of integrating acoustic simulation early in the design process—what they term "parametric acoustics"—to support geometry-performance feedback loops. This strategy is applicable to mycelium-based panels, as confirmed by recent simulation-based research examining how form and layout affect room acoustics. Such workflows allow real-time assessment of acoustic outcomes, helping designers explore configurations compatible with mycelium's moldability and biofabrication constraints.

Mycelium Material Acoustic Performance Assumptions:
Mycelium-based composites have shown excellent sound absorption capabilities, particularly in the mid-to-high frequency range. Due to the lack of physical prototyping in this study, we adopt simulated absorption values (α = 0.6–0.99) based on reliable experimental literature. For example, Gezer and Kuştaş (2024) recorded absorption coefficients as high as 0.99 for mycelium grown on straw, while Walter and Gürsoy (2022) found values exceeding 0.9 for samples cultivated on cardboard waste. Though exact performance depends on substrate, density, and thickness, these studies confirm that mycelium materials can effectively substitute traditional acoustic panels.
Acoustic Simulation Setup in Design Tools:
Once geometric and material parameters are set, acoustic performance can be virtually assessed using simulation tools like Pachyderm Acoustic in Grasshopper. These tools apply ray-tracing and image-source algorithms to estimate reverberation time (RT60) in a reference room. Studies such as Walter & Gürsoy (2023) have used this method to test mycelium panel layouts, integrating lab-measured absorption coefficients into simulations. Room-scale models (~5×4×3 m) allow designers to analyze the acoustic effect of panel variations (e.g., thickness, spacing, or surface corrugation) on speech-frequency bands. As demonstrated by Wiggins (2023), embedding this workflow into the parametric environment supports fast, iterative feedback during design development, enabling acoustic optimization prior to fabrication.

Evaluation and Optimization Strategy:
The final phase evaluates panel designs based on acoustic performance relative to material use. Instead of exhaustive optimization, each configuration is scored using a ratio: reduction in reverberation time (ΔRT60) per unit panel volume. This aligns with prior research in parametric acoustic optimization.
For instance, Vlaun et al. (2016) developed a ray-tracing tool within Grasshopper, coupled to Galapagos, to optimize sound absorption in open-plan offices. They strategically positioned acoustic panels using this system, achieving improved reverberation performance with minimal material effectively balancing acoustic efficacy and resource investment. This real-world example confirms the practicality of using efficient metrics (e.g., RT60 reduction per volume) to guide sustainable design of mycelium panels.

Results & Discussion

As this study is conceptual and simulation-based, results are inferred from literature-supported performance ranges and design logic rather than direct prototyping or physical measurements.

Nevertheless, several key insights emerge from the proposed workflow.

Optimized Geometry Potential: Literature suggests that **corrugated or textured panel surfaces**—especially those with sinusoidal patterns or cavities—can significantly enhance sound absorption in the mid- to high-frequency range (500–1000 Hz), which is critical for speech clarity. Based on simulations from prior studies (Walter & Gürsoy, 2022; Gezer & Kuştaş, 2024), mycelium panels with a surface undulation of 15–25 mm and thickness around 40 mm are likely to achieve absorption coefficients exceeding 0.85 at 1000 Hz. When incorporated into the parametric model, these geometries can be tuned to balance acoustic performance with material efficiency.

Material Efficiency and Layout Strategy: Applying a surface area constraint (e.g., max 60% wall coverage), simulation frameworks (e.g., Pachyderm) in other studies demonstrated that strategic placement—primarily on ceilings and upper wall areas—yielded **up to 50–60% reductions in RT60**, with only moderate use of material. This suggests that **non-uniform distributions**, informed by acoustic hotspots (first reflections), outperform uniform panelling in both effectiveness and efficiency. Conceptually, this supports the viability of a **high absorption-to-material ratio**, guiding sustainable acoustic panel design.

Design Exploration vs. Optimization: Rather than exhaustive optimization, the workflow allows **designers to evaluate trade-offs** by comparing parametric variants across two axes: acoustic benefit (ΔRT60) and mycelium volume. For example, a deeper corrugation might reduce RT60 by 0.4 seconds but increase material use by 30%. A flatter profile might reduce RT60 by only 0.2 seconds but be 50% lighter. Such trade-offs, visualized in a Pareto-like map, allow **qualitative decision-making** for eco-efficient acoustic performance.

Limitations and Assumptions: This study assumes average acoustic properties (e.g., $\alpha = 0.6–0.99$) for mycelium based on current literature. In reality, absorption varies with growth substrate, density, and finishing. Moreover, simulation assumes ideal behavior and omits edge effects, mounting systems, or any potential airflow resistance added by coatings or framing. Still, the results provide a **rational design foundation** for prototyping, and the methodology is adaptable to updated material data.

CONCLUTION

This paper presented a conceptual methodology for optimizing the design of mycelium-based acoustic panels through parametric modelling and performance-driven planning. Drawing on current literature, we outlined a four-stage process encompassing panel geometry parameterization, material performance estimation, acoustic simulation using Grasshopper-based tools, and evaluation via absorption-to-material efficiency.

Our findings underscore the feasibility and potential of mycelium as a viable acoustic material when paired with algorithmic design strategies. Literature-informed simulations indicate that strategically shaped and placed panels can significantly reduce room reverberation time using minimal material, aligning with sustainability goals in architecture.

While the study remains theoretical, it provides a structured workflow for future development. As next steps, physical prototyping, material-specific calibration, and empirical testing are needed to validate and refine simulation assumptions. Ultimately, this methodology opens a pathway for architects to integrate biomaterials like mycelium into acoustically responsive, ecologically mindful design—bridging material innovation with computational form-finding.

REFERENCES

Arenas, J. P., & Crocker, M. J. (2010). Recent trends in porous sound-absorbing materials. *Sound & Vibration*, 44, 12–17.

Arenas, J. P., & Sakagami, K. (2020). Sustainable acoustic materials. *Sustainability*, 12(13), 6540. https://doi.org/10.3390/su12136540

Gezer, E. D., & Kuştaş, S. (2024). Acoustic and thermal properties of mycelium-based insulation materials produced from desilicated wheat straw – Part B. *BioResources*, 19(1), 1348–1364.

Ghazvinian, A., & Gürsoy, B. (2022). Basics of building with mycelium-based bio-composites: A review of built projects and related material research. *Journal of Green Building*, 17(1), 37–69. https://doi.org/10.3992/jgb.17.1.37

Mogu S.r.l. (n.d.). *Mycelium acoustic panels – Technical data sheets*. Retrieved from https://www.mogu.bio

Pelletier, M. G., Holt, G. A., Wanjura, J. D., Bayer, E., & McIntyre, G. (2013). An evaluation study of mycelium-based acoustic absorbers grown on agricultural by-product substrates. *Industrial Crops and Products*, 51, 480–485. https://doi.org/10.1016/j.indcrop.2013.09.017

Pelletier, M. G., Holt, G. A., Wanjura, J. D., Greetham, L., McIntyre, G., Bayer, E., & Kaplan-Bie, J. (2019). Acoustic evaluation of mycological biopolymer, an all-natural closed cell foam alternative. *Industrial Crops and Products*, 139, Article 111533. https://doi.org/10.1016/j.indcrop.2019.111533

Vlaun, N. J. v., van Waart, A., Tenpierik, M., & Turrin, M. (2016). A sound working environment: Optimizing the acoustic properties of open plan workspaces using parametric models. In *Proceedings of SimAUD 2016*.

Walter, N., & Gürsoy, B. (2022). A study on the sound absorption properties of mycelium-based composites cultivated on waste paper-based substrates. *Biomimetics*, 7(3), 100. https://doi.org/10.3390/biomimetics7030100

Authors